CURRENT MOLECULAR BIOLOGY AND BIOTECHNOLOGY PROTOCOLS FOR BIOMEDICAL RESEARCH SCIENTISTS IN CLINICAL MOLECULAR BIOLOGY REFERENCE LABORATORIES.

Philip Ifesinachi Anochie.

CURRENT MOLECULAR BIOLOGY AND BIOTECHNOLOGY PROTOCOLS FOR BIOMEDICAL RESEARCH SCIENTISTS AND OTHER LIFE SCIENCE RESEARCHERS IN CLINICAL MOLECULAR BIOLOGY REFERENCE LABORATORIES.

Philip Ifesinachi Anochie

CURRENT MOLECULAR BIOLOGY AND BIOTECHNOLOGY PROTOCOLS FOR BIOMEDICAL` RESEARCH SCIENTISTS AND OTHER LIFE SCIENCE RESEARCHERS IN CLINICAL MOLECULAR BIOLOGY REFERENCE LABORATORIES.

Published by

Philip Ifesinachi Anochie

Research Scientist/ISID/ESCMID Fellow.

http://www.isid.org/wp-content/uploads/2018/03/2010_ESCMIDFellow_ANOCHIE.pdf

Molecular Biology and Biotechnology Research Group.

Philip Nelson Institute of Medical Research.

INTERNATIONAL EDITION.

CURRENT MOLECULAR BIOLOGY AND BIOTECHNOLOGY PROTOCOLS FOR BIOMEDICAL RESEARCH SCIENTISTS AND OTHER LIFE SCIENCE RESEARCHERS IN CLINICAL MOLECULAR BIOLOGY REFERENCE LABORATORIES.

Published by

Philip Ifesinachi Anochie

Research Scientist/ISID/ESCMID Fellow.

http://www.isid.org/wp-content/uploads/2018/03/2010_ESCMIDFellow_ANOCHIE.pdf

Molecular Biology and Biotechnology   Research Group.

Philip Nelson Institute of Medical Research.

Tel: +2348140624643, +2348173175179, +2348166582414.

E-mail: philipanochie@gmail.com, OR philipanochie@yahoo.co.uk.

ACKNOWLEDGEMENT

The Author gratefully acknowledge the help of God for his support in writing this book.

CURRENT MOLECULAR BIOLOGY AND BIOTECHNOLOGY PROTOCOLS FOR BIOMEDICAL RESEARCH SCIENTISTS AND OTHER LIFE SCIENCE RESEARCHERS IN CLINICAL MOLECULAR BIOLOGY REFERENCE LABORATORIES.

SUMMARY

This book contains a compendium of 34 chapters of full information on current molecular biology and biotechnology principles, protocols, techniques and applications covering such topics such as ; RNA isolation and RT-PCR procedure, Cloning of PCR DNA fragments: Preparation of fragments , dephosphorylation of vector and ligation steps, Transformation of competent cells, Minipreparation and restriction analysis of plasmids from transformed *E. coli.*, Expression and purification of recombinant fusion protein, Detection of protein expression from cloned genes by western blotting, Purification of genomic DNA from corn shoots, Restriction enzyme digestion and gel electrophoresis of DNA and transfer of migrated DNA to a membrane by capillary action (Southern blot setup), Southern blotting: Probe preparation, pre-hybridization, hybridization and development of membranes, DNA sequencing by Dideoxy chain termination method, Plasmid extraction and curing, Trizol preparation of total RNA, genomic DNA preparation, setup of genomic DNA for southern blot analysis, southern and northern blots, basic principles in molecular cloning, transformation, plasmid DNA isolation, sequencing protocol, protein analysis techniques, RNA analysis techniques, RT-PCR assay to amplify cDNA fragment, Gel electrophoresis and cutting out of fragment, genomic DNA extraction, purification of cDNA from agarose, ligation reaction of fragment and pre-cut vector, purification of genomic DNA from corn shoots using proteinase K method, transformation of ligation reactions using the heat shock procedure, restriction enzyme digest and gel electrophoresis of genomic DNA , Capillary transfer of DNA to membrane (Southern Blot), counting of transformants and inoculation of cultures, probe labeling and Southern hybridization, mini- plasmid prep of cultures , and restriction digest of preps to ascertain ligation, identification of positive clones, inoculation of small cultures with recombinant protein expression clone, sequencing reaction and preparation of sequencing gel, expansion of culture to 100 ml preparation and induction of recombinant protein production, running of sequencing gel, purification of recombinant protein using cellulose affinity chromatography, pouring of SDS -PAGE gels, loading of purification sample onto SDS- PAGE , Use of Coomassie Blue stain, western blot analysis, PCR and reverse hybridization assays, use of pipettes and their types, BACTEC ™ MGIT ™ 960 SIRE test for the anti-mycobacterial susceptibility testing of *Mycobacterium tuberculosis* for first and second line drugs, GT-BLOT 20 automated equipment for reverse hybridization assays, SPOLIGOTYPING: a PCR-based method to simultaneously detect and type *Mycobacterium tuberculosis* complex bacteria for epidemiological studies, MIRU-VNTR : mycobacterial interspersed repetitive units -variable number tandem repeats; a strain identification and differentiation protocol for the epidemiological study of *Mycobacterium tuberculosis*, *Mycobacterium tuberculosis* genotyping techniques, Biosafety and laboratory security, good laboratory practices and standards, quality control as well as other safety precautions, rules and regulations in a molecular clinical reference laboratory   and many more.

This book will help you to have a current and up-to- date   clear understanding of the principles, protocols and procedures used in the molecular biology laboratories in   clinical   molecular biology reference laboratories.

CHAPTER 1

INTRODUCTION

Genes, proteins and DNA   are studied in molecular biology and biotechnology. A nineteenth-century monk called Gregor Mendel introduced the notion of genes: basic units responsible for possession and passing on of a single characteristic.

Initially it was thought that proteins carried genetic information until mid 20th-century, when it was found that DNA did.

Proteins are the functional molecules in cells (ie. they perform the majority of the reactions of life) Understanding the basics- DNA: The genetic material in the DNA molecule is made of 2 long spiral, ladder-like strands, called double helix , right handed double helix interact through H_2 bonds .

They are anti-parallel (one oriented 5' - 3', the other 3' - 5') first described by James Watson and Francis Crick in 1953. DNA molecule is made of 2 long spiral, ladder-like strands called double helix.

The right handed double helix interact through H_2 bonds. They are anti-parallel (one oriented 5' - 3', the other 3' - 5'). They were first described by James Watson and Francis Crick in 1953.

Each spiral strand, sugar phosphate backbone and attached bases,

connects to a complementary strand by hydrogen bonding (non-covalent) .

Bonding is between paired bases, purines P, adenine (A) guanine (G)

Pyrimidines P, Thymine (T) and cytosine (C). Adenine pairs with thymine and are connected by two hydrogen bonds (non-covalent) . Guanine and cytosine are connected by three hydrogen bonds.

In the structure of RNA many types of RNA in cells most commonly described are messenger RNA -mRNA, transfer RNA-tRNA and ribosomal RNA - rRNA.

In polymers, they are mostly single stranded with back bone of alternative sugar and phosphate molecules. They have sugar - ribose nitrogen base thymine substituted by uracil.

Nucleotide bases are covalently bonded to sugar molecules and they may fold back on itself to form complementary Hydrogen bonds.

RNA is like DNA but the sugar-phosphate backbone has a different sugar: ribose instead of deoxyribose and where the DNA molecule has the nucleotide thymine (T), RNA has the nucleotide uracil (U).

RNA is almost always a single stranded molecule whereas DNA always stored as a double helix in eukaryotes.

RNA comes in different forms including: Messenger RNA (mRNA) is transcribed from DNA and translated into protein, Transfer RNA (tRNA) is a functional molecule used in the process of translation .

In DNA and RNA preparation and sequencing , we should understand simply that DNA and RNA are nucleic acids. The isolation and purification of nucleic acids is one of the most common tasks in molecular biology research.

The method for isolating and purifying nucleic acids must be selected wisely according to the research application. The type of nucleic acid (ssDNA, dsDNA, total RNA, rRNA, etc.), the source (mammalian, lower eukaryote, plant, prokaryote, or virus), the type of sample (whole organ, tissue, cultured cell, blood, e.t.c), the quality (yield, purity, purification time), and the specific application (PCR, Cloning, Labeling, phosphorylation, blotting, RT-PCR, cDNA synthesis, RNase protection assays, etc.) are all factors that must be considered.

The isolation and purification of nucleic acids is carried out by first breaking down the cell , inactivating the nucleases present in the cell , and finally isolating the nucleic acid.

CHAPTER 2

RNA ISOLATION AND RT–PCR PROCEDURES

In this procedure, you will be using an adaptation of the single step RNA isolation method derived by and patented to Choczynski and Sacchi (1987, *Anal. Biochem.* 162: 156–159). In this book, you will be using the widely used commercially available reagent (Trizol) for RNA isolation. This is a mixture of the RNase inhibitor Guanidium thiocyanate and phenol. Furthermore, we will perform a one step reverse transcriptase polymerase chain reaction experiment using the HotStar Taq Qiagen Kit.

NOTE: It is important to note that RNA is extremely sensitive to degradation . We will know more about this as we go further. Make a special effort to be as aseptic as possible. Wear gloves all the time and change them regularly. We will be isolating RNA from the bacteria *E. coli* . We will be attempting to extract the c DNA sequence of recombinant fusion protein known as " abg– CBD_{cex}. Our overall cloning strategy will be explained below. It is important to note that the polymerase chain reaction is actually an incredible easy procedure to do. General details will be given below.

PROCEDURE:

1. Centrifuge a 1.5 ml suspension bacterial culture at maximum rpms (14,000) in a microcentrifuge for 5 minutes. Make sure your tubes are appropriately balanced (with another group's sample or using additional tubes filled with water to a similar level).

2. Carefully decant or pipette out the supernatant without disturbing the pellet. This should be relatively easy to do since the bacterial pellet is quite sticky. After the supernatant has been removed, use a p200 to add 50 ul of d H_2O and vortex vigorously until the pellet appears resuspended.

3. Add 1ml of Trizol reagent to your pellet and mix by carefully pipetting up and down for a minimum of 20 times with your p1000.
   NOTE: Trizol contains phenol so take care! Incubate the microfuge at room temperature for 5 minutes. Add 200ul of chloroform , shake vigorously for 15 seconds and incubate at room temperature further for about 3 minutes.

4. Centrifuge at 12000 rpm at 4 ºC for 15 minutes. NOTE: After the centrifugation step, you will notice two phases of liquid. The upper phase is aqueous and contains your RNA prep. The lower phase is the organic phase– you don't want this.

5. Transfer the upper phase to a fresh microcentrifuge tube. Take care not to disturb the interphase when retrieving the aqueous phase.

6. Precipitate the RNA by adding 500ul of isopropanol (isopropyl alcohol) , and letting it incubate at room temperature for 3 minutes.

7. Centrifuge at 12000 rpm for 10 minutes at 4 °C . Wash the pellet with 1ml of 70 % ethanol and spin again at 12000 rpm for 5 minutes.

8. Keep the pellet and air–dry for approximately 10 minutes (Do not over dry). Dissolve the pellet in 100ul of DEPC treated distilled water (if the RNA pellet is difficult to dissolve , pass it through a yellow pipette tip a few times or incubate at 55 °C for 10 minutes) . Your samples are now ready for the RT–PCR.

9. The RT–PCR experiment like most kits is usually very straight forward. You will perform three reactions : one control and two reactions using your purified sample – one of which will be diluted. You will prepare the diluted sample by aliquoting 10ul of your RNA sample into a new centrifuge tube. You will then add 90 ul of DEPC treated water thereby diluting your sample 10 fold.

10. The two reactions will be prepared by adding RNA to each of your two Master Mix solutions found in your ice bucket. To master mix #1, you will add 1 ul of your newly purified RNA, and to the master mix #2, you will add 1ul of your diluted RNA. **The Master Mix contains the enzymes (reverse transcriptase and the Hot Star Taq Polymerase) and primers necessary for the reaction. This will be discussed in more detail below.**

11.   The reactions will each require a drop of mineral oil and will then be loaded directly into a thermal cycler machine once an ambient block temperature of 50 °C has been reached.

12.   The thermal cycler program has been set as follows (total time of approximately 3 hours) :

30 min   50°C
15 min   95°C

35 cycles of the following :
1 min   94°C
1 min   60°C
1 min   72°C

Then finally:
10 min   72°C.

This program actually incorporates the reverse transcriptase assay (the first 30 min), a transition period (the next 15 min), and finally the PCR amplification step (the remainder).

13.   Once the reaction has started, you may take spectrophotometry readings of your purified RNA samples. A demonstration will be conducted below.

Dilute 2 _L from above 998_ L of water. Measure the absorbance at 260nm and 280nm , to determine an A_{260}/A_{280} ratio (pure RNA should be about 2.0) and concentration of your RNA prep.

(_g/mL RNA) =(A $_{260}$ nm) x (40 _g/mL) x (dilution factor).

14.   While the reaction is running , prepare the agarose gel and electrophoresis set-up (There will be a quick run through of the equipment used.) You will pour a 0.8 % agarose gel - this is a weight per volume measurement. Weigh out 0.4 g of agarose and transfer to a 100ml flask. Add 50 mls of 1x TBE buffer and swirl gently to disperse the agarose. Microwave the mixture on high power until it boils and the agarose is completely dissolved. Look for the occurence of "chunkies" in your mixture. The dissolving step is a fine line between boiling your sample enough to dissolve your material, but not boiling it too much so that liquid starts to evaporate.

Allow the solution to cool to 60 ºC by incubating in a 60 ºC water bath for about 10 minutes. Then add 1ul of 10mg /ml Ethidium Bromide (EtBr) stock.

WARNING: EtBr is a carcinogen! Do not handle without wearing gloves and avoid spills. While the agarose is cooling off, prepare the gel plates on the casting setup.

(This will be demonstrated).

Pour the cooled agarose into the plate- don't forget the comb! The gel will take approximately 20 minutes to gel.

15. Once the reaction has finished, you will load your samples alongside a RNA control and a molecular weight standard (the lambda HindIII ladder) onto your 0.8 % agarose gel (see legend below). You will then run the gel at 80 V for approximately 30 minutes. Visualize and take a picture of your gel.

To your gel, you will load the following samples:

i. *10ul of a molecular weight standard (10ul of a lambda Hind III standard).*

ii. *To your Master Mix #1, you will add 10 ul of DNA loading dye and load 40 ul.*

iii. *To your Master Mix #2, you will add 10 ul of DNA loading dye and load 40 ul.*

iv. *Finally, you will load 40ul of a cDNA fragment control (provided in the ice bucket and already containing loading buffer).*

16. Cut out the band using a razor blade. Try to take as little excess agarose as you can. Place the agarose chunk (with your band) into a microfuge tube. Store at –20 ° C until the next experiment.

CHAPTER 3

CLONING OF GENOMIC DNA RESTRICTION .

CLONING OF PCR DNA FRAGMENTS: PREPARATION OF FRAGMENTS , DEPHOPHORYLATION OF VECTOR , AND LIGATION STEPS.

DNA cloning is a powerful technique in molecular biology to generate specific recombinant DNA molecules intended for a variety of uses as listed below:

1. To obtain large quantities of specific DNA sequences for use in studying gene structure and gene regulation , DNA/protein sequence determination , and for in vitro mutagenesis and nucleic acid sequence.
2. To produce large quantities of proteins like growth hormones, cell surface receptors, enzymes, etc., for research or commercial use.
3. To modify the host cell's genotype or phenotype.

In this section of this book , we will use the fragment prepared in your RT-PCR assay. We will need to purify this segment from the agarose. Once this is accomplished, we will then perform a ligation reaction using a prepared fragment and a special pre-cut and vector (pCR 2.1) .

We will then detect the presence of successful ligations by transforming bacteria with a ligation mixture. Essentially, only successful ligations (recirculized plasmid, or recircularize plasmid plus insert) will allow colonies to form upon transformation.

It is important to know that you will be using a special gene clean kit to purify your fragment. Other procedures will also work to clean plasmids or to insert DNA but most kits are useful in that they also " melt agarose" for convenient fragment isolation.

PROCEDURE:

1. Estimate the approximate volume of your agarose chunk (you want it to be no more than about 250 ul). Add ½ volume of TBE Modifier and 4.5 volumes of NaI and incubate at 55 °C . This solution should begin to liquefy your agarose sample. Mix by tapping the bottom of the tube without dispersing the solution too much to the sides of the tube.

2. Find the GlassMilk tube in your ice bucket, and mix thoroughly (you may have to dig around with a pipette tip or vortex vigorously) to re-suspend the silica beads. Add 5ul of this to your fragment sample.

3. Incubate on ice for 5 minutes (mix the contents after 2 and 4 minutes of incubation to make sure the beads do not settle to the bottom of the tube).

4. Spin down the beads in the micro-centrifuge for 5 seconds at maximum rpms. Remove the supernatant carefully with a pipet tip making sure not to disturb the pellet which contains the bound DNA.

5. Wash the pellet by adding 700 ul of NEW wash solution . Re-suspend the pellet thoroughly by pipetting back and forth while scraping it off the side of the tube with the pipet tip.

6. Spin down the silica beads for 5 seconds as before. Discard supernatant . Repeat the washing steps two more times. At the

end of the third , remove all the supernatant thoroughly as follows (this is necessary to avoid dilution of the eluting buffer with the residual NEW wash solution).

- Pipette out the supernatant carefully as before.
- Spin down the tube briefly again to bring down ALL the liquid sticking to the sides of the tube.
- Take the last bit of liquid away from the pellet using a thin drawn-out pipet tip.

7. Elute the DNA from the silica matrix by adding 10ul of sterile dH20 and re-suspend the beads thoroughly by drawing back and forth with a pipette tip. Avoid splashing the suspension on the sides of the tube.

8. Incubate the suspension in 55 °C water bath for 3 minutes. Spin down the silica beads in the micro-centrifuge for 30 seconds at maximal speed.

9. Transfer/save the supernatant into a sterile micro-centrifuge tube labeled accordingly. (You can use the drawn-out tips again if you'd like).

10.   To the pellet, add another 10ul of sterile dH20 and re-suspend the pellet thoroughly.

11.   Repeat steps 8 and 9.

12.   Remove the supernatant from the second spin and combine with the supernatant from the first spin (total DNA solution for your sample is now 20ul). Discard the pellet.

13.   You will want to check the fragment from your Gene Clean as follows:

- Transfer 2ul of the purified fragment into a micro-centrifuge tube containing 8ul of dH20 and 2ul gel loading buffer. Load this sample into the "communal " 0.8% agarose gel.
- Once loaded by all participants, the gel will be run for 30 minutes at 100 volts in 1Xtbe.

14. You will then prepare for your ligation reaction which will involve, your cut vector sample (pCR2.1 in the ice bucket) . Label six tubes A to F (or give them descriptive names to help you keep track of what they are. To each tube , add the various ligation components as follows: (numbers are in ul units) . NOTE: Make sure all components are mixed at the bottom of the tube before adding the ligase.

	A	B	C	D	E	F
pCR2.1	1	1	1	1	0	0
Purified fragment	1	2	5	0	0	0
Uncut vector (pEO.1)	0	0	0	0	1	0
5x ligase buffer	2	2	2	2	2	2
dH20	5	4	2	6	6	7
T4 DNA ligase	1	1	1	1	1	1

15. Incubate overnight at 16 ºC and take out your ligation samples the next day.

16. Learn how to prepare buffers and solutions.

CHAPTER 4

TRANSFORMATION OF COMPETENT CELLS

In this module, we will take a look at a tried and true method of introducing DNA into bacterial cells. (Heat shock). Electroporation as an alternative technique will be discussed below. It is important to know that competent cells are very delicate. Be gentle when handling them. Use a competent version of *E. coli* called BL21 (DE3) cells. These cells are commonly used in high yield inducible protein expression systems and will be discussed in more details below.

PROCEDURE:

1. Take your ligation reactions (tubes A to F) from out of the water bath.

2. **Now,** prepare 6 fresh micro-centrifuge tubes and label them from A (t) to F (t). These will be the tubes where the heat shock transformation reactions take place. Keep on ice.

3. Competent cells will be brought out and placed in ice bucket near the start of the laboratory work. Carefully transfer 20ul of competent cells to each of your 6 transformation tubes.

4. To tubes A (t) to F (t) , add 1ul of the ligation mixes prepared since a yesterday's procedure. Leave the cell/DNA mixtures on ice for 30 minutes. Place SOC media in the 37 °C water bath to pre-warm.

5. After the 30 minutes on ice, you will need to heat shock your 6 samples for exactly 60 seconds in the 42 ° C water bath. **Do not agitate cells.**

6. Immediately transfer to ice. Then add 480ul of the pre-warmed SOC media to each tube. Mix gently.

7. Incubate at 37 ° C for about 1 hr with shaking . This will roughly coincide with your electropolated samples.

8. After the incubation step, place 50ul of the contents of each tube onto corresponding labeled LB agar plates containing 100ug/ml ampicillin and 40ug/ml *X-gal.* Spread plate your sample using the turn table and " hockey sticks" . This will be demonstrated below.

9. Incubate all 6 plates overnight at 37 ° C in an inverted position (agar side up). You will take the plates out the following morning.

10. Buffers and solutions are as follows:

LB AGAR; Make 1L of LB broth , and add 15g of Bacto-agar before autoclaving . Cool to 55- 60 ° C before adding ampicillin or X-gal, if necessary.

SOC MEDIUM; 20 g Bacto- tryptone, 5g Bacto- yeast extract, 0.5g NaCl, 950ml distilled H_2O , 10ml of 250 mM KCl, adjust pH to 7.0 with 5M NaOH, adjust volume to 1L, autoclave . Just before use , add 5ml sterile 2M $MgCl_2$ ﹐and 20 ml of filter –sterilized 1 M glucose.

CHAPTER 5

MINIPREPARATION AND RESTRICTION ANALYSIS OF PLASMIDS FROM TRANSFORMED *E. COLI*

The following procedure is based on lysis by boiling procedure of Holmes and Quigley (1981, *Anal. Biochem.* 114:193– 197). The procedure is rapid and convenient for plasmid analysis from transformed cells. The plasmid isolated is usually pure enough for restriction analysis (although can be contaminated with chromosomal DNA), making this protocol , a good alternative to the phenol–chloroform extraction commonly employed.

It is important to know that this is a two day procedure, requiring a short period of time on the first day. You will pick a total of 12 colonies. Remember , you will ultimately want to pick colonies that have your vector + insert combination. Choose your colonies wisely.

Your isopropanol needs to be ice cold when used. Put it in the ice bucket now. If the plasmid prep cannot be done the day following the overnight growth of the culture, the culture can be spun down and the pellet stored at – 70 ° C until needed.

PROCEDURE:

1. (DAY ONE) On the day before the miniprep experiment, you will inoculate 2 mls of LB broth + 50ug/ml ampicillin with a single colony of transformed bacteria. NOTE: don't forget that

you will be inoculating a total of 12 cultures . You may use the sterile toothpicks offered, by dipping the end of the toothpick into the colony and then throwing the entire toothpick into the broth.

2. Grow the culture overnight on the roller drum inside the 37 °C incubator. Remember to label the tubes and balance them properly in the apparatus.

3. (DAY TWO) Vortex each culture thoroughly , and transfer 1 ml of each culture into a clean microfuge tube. Store the rest of the culture in the refrigerator as backup.

4. Spin the cells down for 30 seconds at maximum speed in the microcentrifuge. Remove all the supernatant by pipetting out the last bit of media left.

5. Add 200ul of STET buffer to your cells and re-suspend by vortexing.

6. To your cell mixtures, add 20ul of 20mg/ml lysozyme solution in 25 Mm Tris -HCL , pH8. Vortex briefly, and place tubes in a boiling water bath for 60 seconds . Do not over-boil and don't forget to use a lid-lock.

7. Immediately spin for 10 minutes at maximum rpms in the microcentrifuge.

8. Remove the pellets (predominantly chromosomal DNA) with the flat end of a sterile toothpick without mixing in with the supernatant. Discard the pellet.

9. Add an equal volume of ice-cold isopropanol to the supernatants. Mix. Chill for 10 minutes at -20 ° C.

10.   Spin down DNA precipitates for 5 minutes at maximum speed in the micro-centrifuge. Wash pellet with 250ul of 70% ethanol. Spin down for 1 minute. Remove the supernatant.

11.   Air- dry the pellets or dry in SpeedVac for 10 minutes.

12.   Dissolve the pellet in distilled water (50ul or otherwise depending on size of pellet).

13.   You will now set up 12 restriction digests with the following recipe for each digest:

15ul dH$_2$O

2ul   10x reaction buffer

2ul plasmid prep (actually will depend on pellet size)

1ul restriction endonuclease (to be announced)

Incubate for 1 hour in a 37° C water bath.

NOTE: A common procedure for conducting large numbers of restriction digests is the preparation of a "cocktail" mixture . In this case, you would prepare one main solution containing 8 times (since we have 7 samples) the amount of everything except the DNA (i.e. 120ul of water, 16ul of the react buffer , etc,etc.). This way , to your DNA , you can just add 18ul of this "cocktail " mix. If you are performing digests where you have 18 or 36 samples to deal with, this method can save you a lot of pipetting time.

14.   Prepare a 0.8% Agarose gel as previously outlined in module 3. Do not forget ethidium bromide.

15.   After the digest incubation, add 4ul of DNA loading buffer to each tube and load samples – along with a ^HindIII marker –to your agarose gel. Run gel at 100 V for approximately 1– 2 hours. Visualize and photograph gel.

16.   Discuss possible band profiles. Keep in mind we will be looking for plasmids that contain out fragment of interest in the correct orientation.

17.   Buffers and solution are as follows:

STET

8% sucrose

5% Triton X–100.

50mM Na$_2$EDTA

50 mM Tris –HCL (pH8)

STE

0.1M NaCl

10mM Tris– HCL (pH8)

1mM EDTA.

CHAPTER 6

EXPRESSION AND PURIFICATION OF RECOMBINANT FUSION PROTEIN: DETECTION OF PROTEIN EXPRESSION .

This module highlights common procedures used for the expression and purification of small amounts of recombinant protein. We will be using a system based on cellulose affinity chromatography. This procedure can be easily upscaled to larger production models if necessary.

It is important to know that we have designed this module so that the protein you will be purifying is relatively easy to handle. However, depending on your specific research and the stability of your particular protein , it is wise to be efficient , quick and safe (i.e. keep things cold, use protease inhibitors) when working with proteins. You will be taking small samples during various steps of the procedure.

The colony you start with is designed for the expression of recombinant fusion proteins, and contains a *pGEX* system vector. Here, one is able to express a protein of interest that includes an additional protein domain at its N- terminal side. In this case, the protein domain is called glutathione–S– transferase (or GST for short). The purpose of this additional domain is to allow your fusion protein to be purified easily by affinity chromatography. Presently , the GST system is one of the most popular fusion systems.

We will be using the following clone:

pGEX 2T N32. This will express a *GST–LCK* construct that contains amino acids –8 to 234 of p56lck (actually contains the *N*-terminal , the *SH3* and *SH2* domain of *lck*). (*p56lck* is a tyrosine kinase and details will be discussed below).

PROCEDURE:

1. (DAY ONE) . You will need to pick one of your positive colonies (i.e. bacteria strain containing pCR2.1 vector plus the right insert in correct orientation. We will have back up colonies should you fail to get one in the previous labs) using a sterile toothpick to inoculate a 1ml Luria plus ampicillin culture. Incubate at 37 °C, shaking at –180 rpm , overnight.

2. (DAY TWO) . You will inoculate a larger 100ml Luria plus Ampicillin culture, with your 1ml starter culture. Place in the 37 o C incubator shaking at –180rpm . You will periodically monitor the turbidity of your culture.

3. Turbidity will be accessed using a spectrophotometer with the aim to stop the culture between an O.D.$_{600nm}$ of approximately 0.5 to 0.8 . You may be able to guess when this will occur based on the assumption that healthy *E. coli* double approximately every twenty minutes.

4. Once the O.D.$_{600nm}$ has reached this mark, remove 1.0ml of your culture and place in a microfuge tube. This will be your "Sample before Induction Prep". You will then add 1ml of a 10 mM IPTG solution so that the final IPTG concentration in your culture is approximately 0.1mM. IPTG will serve to induce the

promoter responsible for the transcription of your recombinant protein.

5. Place your induced culture in an incubator of lower temperature (preferably 26 °C or 30 °C) , and leave overnight , shaking at -180rpm.

6. (DAY THREE). Today's procedure should be done where the sample is chilled whenever possible. First, remove 1.0ml of your culture. This will be your "Sample after induction step". Next , spin down the culture at 10,000 g for 20 minutes at 4 °C, and discard the supernatant. To your pellet , add 1.0 ml of Lysis Buffer , and re-suspend sample. You will then transfer your solution to a microcentrifuge tube . Add 20ul of 20mg /ml lysozyme to your solution.

7. Freeze your sample using a pre-prepared dry-ice /ethanol bath (there will be an explanation on this below). Thaw by patiently waiting with your sample in your hands. Repeat the freeze/thaw process at least once more. You want your sample to become snotty which is a good indicator of lysis.

8. Add approximately, 10ug of DNase 1 to your sample and place in a 37 ° C environment . Keep solution at constant agitation until it is no longer viscous (no longer snotty) . Centrifuge sample in a microcentrifuge at maximum speed for 15 minutes.

9. You will notice that your sample will contain two layers. You will have a debris layer at the bottom (pellet), and a clearish liquid supernatant. The pellet is your insoluble fraction , and

the supernatant is your soluble cell lysate. You want the supernatant.

10. Transfer supernatant to a fresh microcentrifuge tube. Using a syringe filter system (demo) , pass the lysate through a low protein binding 0.2um filter and dispense into another fresh microfuge tube. Remove 10 ul of this sample and place in an additional tube . Label this new tube "Sample: soluble fraction".

11. With the remainder (approximately 1ml) of your cell lysate. You will add 50ul of 50% slurry cellulose beads. These beads will have been prewashed in lysis buffer for you and will be kept in your ice bucket. You will then allow your bead plus lysate sample to mix end over end for approximately 2 hrs in a refridgerated environment.

12. Your sample will then be placed in a microcentrifuge and spun at maximum speed for 3 minutes. The beads (hopefully , your protein) will pellet to the bottom of your tube. You will want to save 10ul of your supernatant in a separate microfuge tube and label "Sample: Unbound lysate". Discard the remainder of your supernatant.

13. To your beads, add 1ml of cold lysis buffer. Vortex rigorously for 10 seconds. Spin down culture as above , retaining the beads. Discard the wash supernatant. Repeat this process two more times.

14. To your beads , add a final wash solution of approximately 1.0ml. Mix thoroughly and remove 10ul of the

suspension. This will be placed in a fresh microfuge tube and labeled " Sample : Purified. ".

15.   When you have a moment during today's procedure, you will need to centrifuge your " Sample before Induction Prep'. Do this in the microcentrifuge at maximum speed for 5 minutes. Note the pellet, and remove all supernatant. At the end of this procedure, you will collect all your "Sample " tubes , label them with your name and store in the –20 °C freezer for storage. They will be used in your western procedure.

BUFFERS AND SOLUTIONS:

Lysis Buffer

1.   Triton X–100.
2.   20 Mm Tris pH 7.5.
3.   150 Mm NaCl.
4.   1Mm EDTA
5.   0.025% beta–mercaptoethanol (optional)
6.   0.2m M PMSF.
7.   1.0 ug/ml pepstatin
8.   1.0 ug/ml leupeptin.
9.   1.0 ug/ml aprotinin.

CHAPTER 7

DETECTION OF PROTEIN EXPRESSION FROM CLONED GENES BY WESTERN BLOTTING.

Western blotting of proteins is analogous to Southern Blotting of DNA. Proteins will be fractionated by electrophoresis in denaturing discontinuous polyacrylamide gels (*Laemmli, 1970, Nature. 227. 680-685)*, blotted unto a membrane by an electrotransfer procedure, and probed with a special antibody to the protein of interest. The detection procedure in this case , is a two –step process using a primary antibody which is unlabeled and specific to the protein of interest , and a labeled specific antibody which binds to the constant regions of the primary antibody. The secondary antibody that will be used for this exercise is enzyme – conjugated and will be detected by incubation with a chromogenic substrate.

Essentially, you will be running two polyacrylamide gels with the intent of performing a western blot on one of them. The samples that you will be running are the various samples you saved from your expression and purification protocol. (Chapter 5). In our case, this is a two day procedure, and you will be using that nasty neurotoxin, acrylamide again. When you become familiar with these techniques, it is quite easy to do this in one day.

PROCEDURE:

GEL PART:

1. (DAY ONE): You will first pour two polyacrylamide gels using the BioRad Mini Protean system. There will be a training on how to set up the apparatus.

2. When you have set up the apparatus and are ready to pour the resolving gel section, you will need to prepare a 10 ml solution of the resolving gel using the following recipe (good for one gel):

   5mls resolving buffer

   2.5mls 30% acrylamide /0.8% bis-acrylamide.

   2.5 ml water.

   100 ul 10% ammonium persulfate.

   6.5ul TEMED.

   Don't forget to add the TEMED and ammonium persulfate last , and just prior to gel.

3. This resolving gel will be poured approximately 1 cm below the well line (this will make more sense after viewing the demonstration). Immediately after this step, you will need to pour an overlay using the water saturated butanol.

4. After approximately 10 minutes, the resolving gel should be polymerized enough to allow you to continue pouring the stacking gel . Prepare a 5ml solution of the stacking gel using the following recipe (good for one gel):

   4.5 ml stack buffer.

   0.5ml 30% acrylamide/0.8 % bis acrylamide.

   25ul 10% ammonium persulfate.

5ul TEMED.

Again, don't forget to add the TEMED and ammonium persulfate last!

5. Pour the stacking gel to the brim of the gel cassette and carefully place the comb into the cassette. Do not worry about the slight overflow of acrylamide. The stack will need about 1hr to polymerize fully, but we will allow our gel to polymerize overnight at 4 °C.

6. (DAY TWO) : Take your gels out of the fridge and allow them to warm up to room temperature.

    (a) During this time, you will begin to prepare your samples. You will have the microfuge tubes labeled:

    i.   Sample before Induction Prep. (pellet).

    ii.   Sample after Induction Prep. (pellet).

    iii. Sample Soluble Fraction (10ul).

    iv.   Sample unbound lysate   (10ul)

    v.   Sample purified (10ul)

(b) To each tube , add 50ul of a 1x sample buffer (blue stuff) to each tube . Just before the hour of polymerization is up, take your five samples and your prestain standards ("STD" tube with pink liquid– does not need sample buffer), and boil them for a minimum of 8 minutes.

7. Whilst your samples are boiling, prepare the gel set up (this will be demonstrated), such that (with the electrophoresis

buffer) the upper buffer chamber is full, and the lower buffer chamber is filled up at least an inch over the bottom of the gel- again, means nothing until you see the demonstration.

8. You are now ready to load your samples. You can load using your p20 pipetteman and yellow tips (you can also use the thin drawn-out tips provided). In general, place the tip directly into the well and slowly push the liquid out , taking care not to introduce bubbles. You can even use the same tip throughout the loading procedure if you rinse the tip out in the upper buffer chamber between samples. Essentially, the following lane order is a guideline and applies to both gels (remember that one will be stained for total protein , and the other will be used for western analysis. NOTE: You will load most sample for the gel destined for the protein stain since it is much less sensitive than the western blot procedure)

Lane 1   2   3   4   5   6   7   8   9   10

9. Put the lid on the gel apparatus (stay colour coordinated) and set voltage to 100 V. The gel will probably take about 1 ½ hours to run. You want to stop it when the dye has reached (but not past) the bottom of the gel.

10.   Whilst the gel is running , prepare your transfer buffer by simply taking the supplied "transfer buffer" and adding methanol until it is approximately 20%– 25% methanol (this may be already done for you). About 10 minutes before the gel is ready , you will also need to prepare your PVDF membrane

(immubulon P) by prewetting in 100% methanol for a few seconds in a small plastic container. Dump out the methanol (down the sink0, and add a small volume of transfer buffer (+20 % methanol) to cover the membrane. Let the membrane soak until transfer procedure is ready. NOTE: At this point, one of your gels will be used to begin the transfer procedure , while the other will be stained in the following manner.

- Carefully place the acrylamide gel into a plastic container filled with coomassie solution (you only need enough stain to cover the gel) . Let the gel incubate, rocking, for a minimum of 20 minutes. After 20 minutes, pour the stain back into a special container marked "used stain' . Add destain solution to your gel and incubate for 20 minutes to overnight. Repeat until the stain has gone from the gel.

TRANSFER PART:

11.   When you set up your transfer, you need to think a little. Basically, you want the proteins in your gel to migrate onto the PVDF membrane. Sounds simple , but inevitably you may one day accidentally mix things up, and your proteins will run away into the buffer.

12.   Remember , your proteins are coated with SDS so they are essentially negatively charged. Therefore, they will move towards the positive electrode , away from the negative electrode. BIORAD has been clever enough to make sure all their transfer systems are colour coordinated. The plastic transfer cassette ALWAYS has a black side. This does not

represent evil. This represents negative charge: your proteins will migrate away from the black side.

So, when you set up the transfer, do so in the following manner (this will be demonstrated). Put gloves on.

(a) In the large plastic container, place your plastic transfer cassette with the black side flat. Add transfer /methanol buffer to submerge the cassette. Take one of the prewetted sponges (rinse in tap and then in distilled water), place on top of black side. In subsequent steps, make sure everything is submerged in transfer buffer.

(b) Add two pieces of Whatman filter paper on top of the sponge. Place your gel next. AND THEN place your membrane on top of the gel. Place two more pieces of Whatman filter paper on top and make sure you get rid of bubbles in between this sandwiched set- up. Place other prewetted sponge on top of all of this , and close the cassette.

(c) Place cassette in transfer holder (remember to make sure everything is colour coordinated) . Add ice holder. Fill chamber up with transfer/ methanol buffer (you may have to use the stuff in your big plastic container) . Plonk the whole thing in your ice bucket. Put the lid on and set voltage at 100V. The transfer will take approximately 1 hour. During this time, it is appreciated if you wash the plates and gel apparatus with tap water rinsed with distilled water in the sink.

13.	When an hour has passed , turn off the power supply, and carefully remove the membrane with a pair of tweezers. Place membrane, protein side up on a clean paper towel to dry for a minimum of 1 hour.

WESTERN PART: (QUICK METHOD DEVISED BY MILLIPORE FOR USE WITH IMMOBULON P MEMBRANE)

14.	After drying, add your primary antibody solution (10 mls of anti-p56lck rabbit antisera "54-3b" at 1/2500 dilution in TBS +Tween 20 and 5% BSA) to your membrane in a clean plastic container. Mix solution around so that the membrane is completely immersed. Place on shaker for about 1 hour. You may find that your membrane looks like its half wet and half dry– this is normal so don't fret.

15.	Pour back the antibody solution into the 15ml FALCON tube. This antibody solution can probably be used 3 more times. Wash your membrane by addition of approximately 20 mls (you can measure it the first time , and eyeball it from that point on) TBS + Tween . Shake by hand for about 10 seconds and dump the solution into the sink.

16.	Add your secondary antibody solution to your membrane (10ml Goat anti-rabbit IgG heavy and light w/alkaline phosphatase conjugate @1/5000 dilution in TBS +Tween 20 +5% BSA.). Incubate on shaker for about 30 minutes.

17.   Pour back secondary antibody solution back into 15 ml tube. Wash membrane as above for at least 6 to 8 washes. Wash once in buffer that does not have Tween 20 detergent (we will use 0.1 M Tris pH 9.5). Your membrane is now ready for substrate detection protocol which will be outlined (we will be using the ASBI– Fast Red).

BUFFERS AND SOLUTIONS.

Resolving buffer:

0.75 M Tris– base.

0.21 % SDS

Adjust to pH8.8 with HCL.

Stacking buffer:

0.13 M Tris– base.

0.12 % SDS.

Adjust to pH 6.8 with HCL.

3x Sample Loading Buffer:

150Mm Tris pH 6.8.

6mM Na_2 EDTA.

3% SDS.

3% b–mercaptoethanol

24% glycerol.

Speck of bromophenol blue.

10x Electrophoresis Buffer:

30g Tris Base.

144g Glycine

10g SDS

Add dH_2O to 1 L

Coomassie Blue Stain.

3.75g Coomassie Brilliant Blue R-250.

750 ML Methanol.

600 ml dH_2O

Stir 2 to 3 hours. Filter through Whatman#1 filter paper.

Destain solution.

5% Methanol.

10% acetic acid

85% dH_2O

TBS (Tris Buffered Saline)

50mM Tris pH 7.5

150 mM NACL.

(+0.05% Tween 20 for TBS = Tween 20)

Alkaline Phosphate Buffer

1.0   M Tris pH9.5

100 Mm NaCl.

5mM $MgCl_2$.

CHAPTER 8

PURIFICATION OF GENOMIC DNA FROM CORN SHOOTS

The following procedure is modified from the Proteinase K digestion /phenol –chloroform purification process by Strauss (1987,pub in Current Protocols in Molecular Biology . *Ausubel et. al. eds.* pp 2.2.1– 2.2.3), which is still in routine use for the preparation of genomic material from many organisms. In this case, DNA will be isolated from corn shoots by lysis in Sarkosyl and proteinase K followed by phenol–chloroform extraction and ethanol precipitation . The DNA isolated by this procedure is sufficiently pure for restriction enzyme digestion.

NOTE; Phenol– chloroform is bad for you. Note handling procedures, we are most interested in genomic DNA which can shear very easily. Although, we are not using strict techniques to prevent shearing . We want to be relatively gentle with our samples today.

PROCEDURE:

1. Harvest 1 gram of shoot tissue from the seedlings provided. Do this as quick as you can in other to keep things cold. Add liquid nitrogen to cover the shoots, and grind the sample to a powder as rapidly as you can. Keep the sample frozen through out the process.

2. Transfer the powder to a 50ml screwcap Nalgene Teflon tube. This is located in your ice bucket with 4.5ml of cold Digestion Buffer.
   Make sure the powdered tissue is thoroughly dispersed in the buffer by gently swirling the tube each time the sample is added.

3. Add 40ul of Proteinase K (25mg/ ml stock solution) to give a final concentration of 0.2mg/ml. Mix gently. Add 0.5ml of 10% sarkosyl to give a final concentration of 1%. Again, mix the samples very gently.

4. Incubate the mixture at 55 ° C with gentle shaking for about 2 hours (generally people will do this step for much longer).

5. Take the sample to the fumehood. With gloved hands, add an equal volume   of phenol mixture (approximately 5ml). CAUTION: Phenol is toxic and extremely caustic. Avoid contact with skin. Open containers and dispense solutions of phenol-chloroform in the fumehood only!.
   Phenol is also buffer saturated , so take a close look at the container. You will notice that there may be two layers of liquid

. The bottom layer is the organic phenol layer . Cap the tube securely and mix the contents thoroughly but gently by inversion until the mixture is homogenous.

6. Centrifuge at 1000 rpm in JA −14 rotor for 10 minutes. Transfer the aqueous phase (top layer) into a fresh tube with a siliconized sterile Pasteur pipet taking care not to disturb the interphase. Repeat the extraction of the aqueous phase with phenol mixture once more (or until the interphase looks relatively clear). Transfer the last aqueous phase to a 15ml screw cap conical FALCON tube.

7. Add an equal volume of a 24:1 mixture of chloroform: isoamyl alcohol. Mix thoroughly but gently. Centrifuge in the tabletop BECKMAN centrifuge for 5 minutes at 3000 rpm.

8. Transfer the aqueous phase to a sterile siliconized 50 ml FALCON tube using a sterile siliconized Pasteur pipet. NOTE: Take care not to disturb the interphase at this step. In your final aqueous sample, you do not want any organic material to carry over.

9. Determine the volume of the aqueous phase using an empty glass pipette . Add 0.5 volume of 7.5 M NH_4 OAc. Mix briefly, then add 2 volumes of 100% ethanol. Cover the tube securely with parafilm and mix the contents gently by inversion. A white stringy precipitate should appear.

10. Prepare a sealed Pasteur pipette (do not use siliconized Pasteur pipettes) using a Bunsen burner (there will be a quick demonstration), and allow the tip to cool. Once cool (wait a

minute or two), you will use this pipette to carefully spool out your DNA.

11.   Dip the spooled DNA into a solution of 70% ethanol. You can put 1:0 ml of 70% ethanol in a microfuge tube. Carefully swirl the pipette tip in the solution taking care not to dislodge your DNA. NOTE: The 70% wash will not dissolve the DNA pellet but will remove excess salts.

12.   Place the Pasteur pipette with the tip facing up for approximately 10 minutes. This will allow our DNA pellet to dry so that all trace of ethanol has evaporated.

13.   Dip the pipette tip in 1ml TE (PH8) and re-suspend until your DNA pellet has been dislodged from the Pasteur pipette and is now sitting in your solution. NOTE: You may need to use a sterile yellow tip to help dislodge the DNA. Incubate at 55 °C until pellet is dissolved. Will take a minimum of 1 hour , but we will incubate for a full overnight step.

BUFFERS AND SOLUTIONS.

Digestion Buffer.

100mM NaCl

10mM Tris-HCL, pH8.

25mM EDTA , pH8

Added after resuspending cells:

1% Sarkosyl and 0.2mg/ml Proteinase K.

TE (pH8)

10 mM Tris– HCl, pH8

1mM Na_2EDTA

Phenol – Chloroform Mixture.

25 parts buffer equilibrated phenol.

25 parts chloroform.

1 part isoamyl alcohol.

CHAPTER 9

RESTRICTION ENZYME DIGESTION AND GEL ELECTROPHORESIS OF
DNA , TRANSFER OF MIGRATED DNA TO A MEMBRANE BY CAPILLARY
ACTION (SOUTHERN BLOT SET-UP).

A large number of restriction enzymes are now commercially
available for use in DNA manipulations. All require Mg^{2+} as
cofactors. The optimum activity for these enzymes largely depend
upon the salt concentration and pH of the buffer solution and the
temperature at which the digestions are carried out. It is important
to consult the manufacturer's recommendations for the use of each
enzyme.

Under optimum buffer and temperature conditions for enzyme activity, digestibility of a DNA sample depends on the characteristics of the DNA substrate itself (i.e. purity, conformation, and degree of methylation (Gruenbaum *et. al.* 1981, Nature 292: 860– 862). In most cases, problems arise mainly from contamination with proteins , polysaccharides or organic solvents during the isolation process.

Gel electrophoresis of our DNA enables us to separate DNA on the basis of molecular weight . At a neutral pH , DNA molecules have a relatively constant negative charge density , and as long as we can ensure identical conformation of our DNA molecules (i.e. by restriction digesting which cuts our DNA molecules into a linear conformation), the migration rate of a double stranded DNA fragment will be inversely proportional to its fragment size.

NOTE: This procedure is pretty straightforward. However , you will be dealing with very small volumes of liquid which means that mixing everything together can be a bit tricky. The easiest way to mix is to just put all the constituents in your microfuge tube and then give the tube a very quick centrifuge spin (often called a "buzz"and lasting only about a second or two – don't forget your balance). Don't forget that your enzymes are worth their weight in gold (at least). Depending on the laboratory work in, there will invariably be some sort of protocol to ensure that the enzyme stock lasts a long time now.

PROCEDURE:

1. Using a spectrophotometer, determine your DNA concentration and yield. There should be a small demonstration on how to use a spectrophotometer (in our case , it is Pharmacia's GeneQuant).

2. Transfer the DNA solution with a P1000 micropipet to a microcentrifuge tube properly labeled for storage.

3. Take a 5.0ul aliquot of the DNA solution using a pipet tip , cut off 4mm from the tip (this is to make the hole bigger at the tip– since we are isolating genomic DNA , we want to minimize shear forces caused by forcing our sample through a small pipet tip size). This small sample of DNA we will add to 495ul of TE (pH8). Freeze the rest of the original DNA solution in –20 °C. Mix the diluted DNA sample (basically a 100 fold dilution) thoroughly. This will be used to take our spec readings.

4. Determine the absorbance for your diluted sample at 260 nm and at 280nm. Calculate the A_{260}/A_{280} ratio to assess the purity of your sample (pure sample should have an A_{260}/A_{280} ratio approximately 1.8). Also calculate the DNA concentration of your original DNA sample (the one you have put in the –20 °C) using the following formula: (ug/ml DNA) = (A260 absorbance) x (50ug/ml) x (dilution factor).

5. To set up your digests, you will need to label 7 microcentrifuge tubes. Add the restriction digestion components in the following order. NOTE: Don't forget to cut your pipet tip by about 4mm when aliquoting the genomic DNA samples– again to prevent shearing.

TUBE Enzyme (10 U/ul)	H20 (ul)	10x BUFFER	DNA (ul)
1.		x	2ul REACT#3
y		1uL (BamH1)	
2.		x	2ul REACT#3
y		1uL (EcoR1)	
3.		x	2ul REACT #3
y		1ul EcoR1+1ul Bam H1	
4.		x	0
y		Undigested	
5.	7		2ul REACT #3
10		1ul (Bam H1)	
6.	7		2ul REACT #3
10		1ul (EcoR1)	
7.	7		2ul REACT #3
10		1ul EcoR1+1ul Bam H1	

NOTE: In tubes 1 to 4 , you will be using your DNA sample (y). You will need 2ul of DNA per reaction (refer to your calculations from the previous week). You want a total reaction volume of 20ul – adjust the water volume accordingly. In tubes 5 to 7, you will be using genomic corn control.

6. Store tube #4 on ice until gel electrophoresis. Incubate the rest of the tubes in a 37 ° C water bath for a minimum of 1 hour. At this point, you will pour an agarose gel.

While the DNA samples are incubating with restriction enzymes, prepare the agarose gel and electrophoresis set up(There will be a quick run through the equipment used). You will pour a 0.8% agarose gel- this is a weight per volume measurement. Weigh out 0.4g of agarose and transfer to a 100ml flask. Add 50mls of 1x TBE buffer and swirl gently to disperse the agarose. Microwave the mixture on high power until it boils and the agarose is completely dissolved. Look for the occurence   of "chunkies" in your mixture. The dissolving step is a fine line between boiling your sample enough to dissolve your material , but not boiling it too much so that liquid starts to evaporate. Allow the solution to cool to 60 °C by incubating in a 60 °C water bath for about 10 minutes.

Then add 1ul of 10mg/ml Ethidium Bromide stock.

WARNING: EtBr is a carcinogen! Do not handle without wearing gloves and avoid spills.

While the agarose is cooling off , prepare the gel plates on the casting setup. (This will be demonstrated). Pour the cooled agarose into the plate- don't forget the comb! The gel will take approximately 20 minutes to set.

7. When your restriction digest incubations are finished , add 4ul of blue loading dye to all seven tubes. Buzz your samples in the microfuge if necessary.

8. Add 1x TBE electrophoresis buffer to the chamber to a depth of about 3mm above the surface of the gel.

9. Load your seven samples (now blue) into the gel. There will be a gel loading demonstration. Make sure you keep track of what lane is what . Include a ^-HindIII marker as your eight sample (correct amount will be announced) – this sample will be your molecular weight marker.

10.   Cover the electrophoresis chamber carefully so as not to disturb the samples. Attach power cords to the appropriate terminals of the gel chamber and power supply.

11.   Turn on the power supply and set to 100 V (constant voltage) . You will need to run the gel until the dye is close to the bottom of the gel (this will take about 1 to 2 hours).

12.   When the gel is ready, turn off the water supply , and carefully transfer the gel to the UV light box. Since we added ethidium bromide during the gel preparation, it is already ready to visualize. Wearing protective glasses, look at your gel, and take a polaroid of the data.

13.   Carefully transfer the gel into a clean plastic tray in preparation for the southern blot. Clean the UV lightbox by spraying with 95% ethanol and wiping with a kimwipe tissue.

14.   To the plastic container, add enough 0.25M HCl to cover the gel. Incubate at room temperature on the rocking platform

for about 20 minutes or until the bromophenol blue marker on the gel turns yellow. While the gel is incubating in the HCl, cut a piece of Hybond N+ filter to the size of the gel. Label a corner of the membrane with your initials using a pencil. NOTE: The membrane we are using today is a positively charged membrane. This actually allows us to use a quick method of capillary transfer.

15.   Pour off the HCl carefully. Rinse the gel briefly in distilled water. Replace the water with 0.4M NaOH and leave the gel in this solution while the transfer set-up is being prepared. NOTE: Make a demonstration.

16.   Let transfer occur overnight. Tomorrow , we will dismantle the setup and vacuum, bake your membrane at 80 ° C for 2 hours. Now your membrane is ready for probing.

BUFFERS AND SOLUTIONS:

Loading /Stop Buffer

50% Glycerol.

0.1M EDTA.

1% SDS (optional)

0.1% bromophenol blue.

10x TBE Electrophoresis Buffer.

108g Tris base.

55g Boric acid.

40ml of 0.5M EDTA (pH8)

Water to 1L.

CHAPTER 10

SOUTHERN BLOTTING: PROBE PREPARATION , PREHYBRIDIZATION, HYBRIDIZATION, AND DEVELOPMENT OF MEMBRANES.

We will first prepare a labeled probe which can be used on our membranes. This probe is derived from 28s ribosomal sequences from soybean and is called pGMR1. In general, all of these techniques that have been termed as blotting techniques utilize a procedure for immobilizing macromolecules from gels onto a solid support. This way, manipulation of our material is much more convenient than dealing with a porous gel matrix. This week's procedure will require you to come in two consecutive days: one day where you will prepare your probe and incubate your membrane in prehybridization and hybridization solutions, and one day where you will wash and develop your membrane to visualize your results.

NOTE: We will be using a biotinylation and chemiluminescence detection system this week.

PROCEDURE:

1. (DAY ONE) . Today, we will be using hybridization bottles (tube rollers) to incubate our membranes. There will be first be a quick demo on how to set these up.
2. Obtain a 25ml solution of your prehybridization buffer (Add this solution to your membrane in the hybridization bottle.)
3. Denature 500ul Herring Sperm DNA (10mg/ml) in a boiling water bath for 5 minutes. Immediately chill on iceand add to your prehybridization buffer.
4. Incubate your prehybridization step in the rotating hybridization oven for a minimum of 2 hours at 42 °C.
5. At this point, we will go into PROBE LABELING MODE, and prepare our probe. NOTE: We will be using non-radioactive biotin-14-Datp labeling to allow visualization of our probe. This system has high affinity for strept-avidin which can be complexed with an enzymatic means of detection (in our case , we will be using a Strept- Avidin-Alkaline phosphatase conjugate).
LABELING REACTION: For each NICK translation reaction, pipet the following components (all included in the kit except the DNA sample) into a 1.5 ml microcentrifuge sitting on ice:
5ul 10x dNTP mix.
x ul DNA (equivalent to 1ug) (to be announced).
y ul sterile dH2O to 45ul.
Mix briefly by tapping the tip of the tube. Add 5ul 10x ENZYME MIX. Cap the tube and mix gently. Spin down for 2 seconds in the microcentrifuge using the momentary spin button.

Incubate in the 16 °C water bath for 1 hour. Add 5 ul stop buffer to the reaction mix.

PREPARATION OF THE NICK COLUMN:

Set up your stand such that you can place the column conveniently over a set of tubes (which you will use to collect the eluent) . Take your column , remove the cap, and cut the bottom tip to allow liquid to drip out. You will want to flush the column a minimum of three times with 1ml of TE buffer. Do not worry about too much liquid dripping out such that the column dries out – the column should have a safety filter at the top which will prevent this from happening.

–Load your biotinylated sample onto the column. Wait for it to enter the column matrix. Then add 400ul of TE buffer and collect the eluent (this eluent will be waste but treat as if it is radioactive). Once the first 400ul has passed through , position a clean microcentrifuge tube under the column and add another 400ul of TE into the column. The eluent that comes out at this step will be your labeled probe.

6. Just before the prehybridization incubation step is finished, denature all of your biotin–labelled probe in a boiling waterbath for 5 minutes. Go to your prehybridization tube and empty its contents. Add 25ml of hybridization buffer and add the denatured probe to your bottle.

NOTE: The amount of probe added to the hybridization bottle may change during class (want about 100ng/ml)– keep an ear out for the pertinent details. Also, you will need to add a previously prepared labeled probe that picks up the molecular weight standards used. (^HindIII).

7.Incubate overnight at 42 °C in the rotating hybridization oven.

8. (DAY TWO) Discard the hybridization buffer down a sink and perform the following wash steps.

9. In the hybridization bottle , perform the following wash steps:

i. 2x SSC, 0.1%SDS for 3 minutes, room temp and repeat once more.

ii. 0.2x SSC , 0.1% SDS for 3 minutes, room temp and repeat once more.

iii. 0.16 x SSC 0.1% SDS for 15 minutes at 65 °C, repeat once.

iv. 2 x SSC , for 3 minutes , room temprature and repeat once more.

vi. Expose the film overnight at - 70 °C.

10. You will then allow binding of SA- AP to the Biotin- Labeled Probe in the following manner.

11. Rinse the membrane in Buffer 1 for 1 minute.

12. Replace buffer 1 with freshly prepared Buffer 2 prewarmed to 65 °C. Incubate at 65 °C for 1 hour.

13. We will then hand out a 10 ml SA-AP solution , which has been freshly prepared by spinning down the SA-AP conjugate, and added to 10 ml of buffer 1 to a final concentration of 1ug/ml. Add this solution to your

membrane and incubate on a rocking platform at room temperature for 10 minutes.

14. Membrane will then be treated with GibcoBRL PhotoGene Chemilluminesence system and placed on autorad film (DEMO).

BUFFERS AND SOLUTIONS:

20 x SSC (pH7)

3M NaCl

0.3 M Na Citrate.

Autoclave.

50x Denhardt's Solution.

1% (w/v) Bovine Serum Albumin (BSA).

1% (w/v) Ficoll

1% (w/v) Polyvinylpyrrolidone.

Filter –sterilized , store at –20 ºC.

Prehybridization Solution.

50% (v/v) formamide.

5X SSC

5X Denhardt's

Hybridization Solution.

50% (v/v) formamide.

5X SSC

5X Denhardt's

Buffer 1

0.1 M Tris HCl pH 7.5
    0.15M NaCl.

Buffer 2

3% (w/v) BSA in Buffer 1

Buffer 3

0.1 M Tris HCl Ph9.5.
0.15 M NaCl
50mM MgCl$_2$.

CHAPTER 11

DNA SEQUENCING BY DIDEOXY CHAIN TERMINATION METHOD

This procedure is the industry standard for sequencing, and was developed by Sanger , Nichlen and Coulson (1977, *Proc. Natl. Acad. Sci. U.S.A* . 74. 5463–5468). Although the majority of labs now use automated sequencing services, this is still a useful technique to know should you require to do it yourself. We will begin sequencing using a single a single stranded template . We will discuss various types of template required for sequencing reactions in this chapter.

NOTE: Again , we will be using radioactivity today. Our isotype for this procedure is ^{35}S–ATP, a much weaker isotope than 32 P, but still requiring the upmost in caution when in use. Make sure when in use , you keep a mental track of what is "hot" and whether it is behind some sort of shielding. Also make sure you keep tabs on your whereabouts and the whereabouts of anything that may be radioactive. In the event of a spill, inform the instructor immediately. All these precautions are just common sense but they serve to protect you and your laboratory workers.

This procedure is short, however, the sequencing protocol will continue into a second week where you will need to come in for 2 days. Note that due to radioactive isotope usage rules, we may be using an alternate but similar technique.

PROCEDURE:

1. **(DAY ONE)**. You will be supplied a sample of single stranded template that was derived from M13 cultures (phage procedures) . You will be trying to get sequence from this piece of DNA.

2. **Anealing:** The first step in the preparation of sequencing reactions is the annealing of the primer to the template DNA . To do this , mix 7ul of template DNA , 1ul of primer (0.5 pmole /ul) and 2ul of 5x sequenase buffer in a microcentrifuge tube

and incubate at 65 º C for 2 minutes (this particular step, we will try to get the entire laboratory staff to incubate at the same time). Then place the tube (s) in a beaker of 65 ºC water and allow it to cool slowly to 30 ºC. During the cooling step , you can prepare the remaining reagents for the reactions described below. NOTE: We want a slow cooling step which can take as long as 1 $^{1/2}$ hours.

3. **Labeling:** Prepare a dilution of the labeling mix (supplied in the kit) by mixing 1ul of labeling mix with 4ul of sterile distilled water in a microcentrifuge tube. Place this mixture on ice until needed. When your annealing reaction has cooled to 30 ºC, begin the labeling reaction by adding (to your template/primer), 1ul of 0.1M DTT and 2ul of the diluted labeling mix. We will then supply 0.5ul of 35 S-dATP, and 2ul of diluted sequenase enzyme (previously diluted 8 fold).

   Mix the contents of the tube (being careful not to introduce bubbles). Allow labeling to proceed by incubating at room temperature for 2 to 5 minutes.

4. **Termination:** In your ice bucket, there should be 4 tubes labeled C, G, T and A. These tubes contain 2.5ul of the ddNTP termination mixes, i.e. 2.5ul of ddGTP mix in the tube G, and etc. Prewarm these tubes at least 1 minute prior to adding the labeling reaction. At the end of the labeling reaction (step 3), add 3.5ul of the labeling reaction to each of the four terminating tubes. Mix each termination reaction reaction with the pipet tip (pipet up and down gently - no bubbles!).

Incubate your 4 termination tubes at 37 °C for a further 3 to 5 minutes. Stop the reactions by adding 4ul of the formamide-dye solution supplied with the kit. Place tubes in the –20 °C freezer until next experiment.

5. **Preparation of a sequencing gel sandwich** : This procedure will take about 2 hours and will be demonstrated. First you should prepare the gel solution since the dissolving step could take as long as 45 minutes. Prepare an 80 ml solution in a 250ml Erlenmeyer flask by mixing 40.4g of urea, 31.2ml of distilled H_2O, 12.6ml of a 38% acrylamide : 2% bis–acrylamide solution and 8.0ml of 10x electrophoresis buffer.

Acrylamide is a dangerous , toxic chemical – it is a neurotoxin. Wear gloves, a lab coat and eye protection. Dissolve the urea by incubating in the 42 ° C incubator. To prepare a sequencing gel sandwich , wash one long and one short glass plate, one set of spacers (2) and two sharktooth combs thoroughly with liquid dishwashing detergent. Rinse all of the items with distilled water and allow them to air dry. Wash one side of each glass plate with 95% ethanol and rub the plates with a kimwipe towel to clean them. Siliconize the long plate clean side by wiping 4 drops (one in each corner) of "Sigmacote" with a Kimwipe. Place the long plate (siliconized side up) on a test tube rack on a bench and lay the spacers along the sides (you can squirt the spacers with a bit of water to make them adhere better to the glass). Place the short side (clean side down) on top of the long plate making sure the spacers stay neatly at the sides. Seal the plates together by taping the bottom and sides with one continuous piece of tape. The top is where the plates don't align because of their different sizes . Place clamps along the sides (i.e. over the spacers; do not clamp the bottom).

After the urea has dissolved , filter the solution using a Buchner funnel, filter paper and a 250ml suction flask.

MAKE SURE EVERYTHING IS READY BEFORE PROCEEDING TO THE NEXT STEPS: While wearing gloves, a lab coat and eye protection , add 0.7ml of a freshly prepared 10% solution of ammonium persulfate to the gel solution, mix by swirling and add 2.5ul of TEMED (again mix by swirling). The addition of these two components will begin the polymerization of your gel (you have about 10 minutes).

To pour the gel , follow the instructions given to you in the demonstration.

Your main objective , is to get the solution into the glass sandwich in as little time as possible without introducing any bubbles. As the solution nears the top of the sandwich , lower the plates to a near horizontal position by resting the top end on a glass pipette. The sandwich should be completely filled to the top with plenty of extra liquid solution overflowing the top. Insert the flat sides of the two shark tooth combs between the glass plates to a depth of about 4mm. Place a wet paper towel near the top of the gel (make sure it doesn't actually touch the gel, or else capillary action will suck the acrylamide out into the paper towel). Cover the gel and wet paper towel with saran wrap and let it polymerize overnight. (The saran wrap and paper towel will prevent the gel from drying up overnight).

6. **(DAY TWO). Today we will be running the gel:** There will be a demonstration. Begin a pre-electrophoresis run which allows the gel to warm up to a uniform temperature. Remove the tape from the sides and bottom of the gel sandwich and remove the combs from the top by gently sliding them out. Wipe the sides of the glass plates with a Kimwipe making sure the short side in particular is spotless (it will rest against the metal plate on the gel apparatus and needs to be cleaned so that the warming step works properly. This plate should distribute heat evenly across the gel . If any local differences in temperature occur because of dirt or water droplets , your gel may result in smiley looking bands.)

7. Place the gel into the electrophoresis box with the short plate on the inside. The box has built in clamps for holding the glass plates in place and a rubber gasket at the top to prevent the upper reservoir buffer from leaking down between the inner plate and the metal plate.

8. Add 1x electrophoresis buffer to the upper reservoir, check for leaks around the rubber gasket and rinse the flat upper surface of the gel with buffer (you can use a syringe). Add buffer to the

lower reservoir, close the reservoir covers and connect the gel box to the power supply. Turn on the power supply and set the power at 55 watts (constant power).

NOTE: It is important to always turn the power off when working with the gel (ie. Loading samples) because the electrophoresis conditions employ a dangerous high voltage. Run the gel for 1 hour ; the front glass plate will become warm to the touch.

9. Loading and running the polyacrylamide gel: When the gel has pre-run for 1 hour, then the sequencing reactions are ready for loading. With the power off, again rinse the top part of the gel with a syringe and 1x electrophoresis buffer. The sharktooth combs are used to form wells (or lanes) for the samples. The side of the comb with the pointed teeth is gently inserted between the glass plates until their teeth just contact the surface of the gel and cause a slight indentation (no more than 2mm into the gel). Once the combs are in place, do not move them again and take care not to bump them during sample applications. Rinse the surface of the gel between the teeth of the combs immediately before loading samples to remove urea.

Heat the sequencing reaction (from last week experiment) at 75 o C for 2 minutes and spin them down briefly. Use a p20 pipetteman and a narrow pipette tip to load 2 to 3ul of the sequencing reaction samples between the tip of the comb. Place the pipete tip between the glass plates and down as close as possible to the surface of the gel before slowly dispensing the sample. Avoid generating air bubbles! Load the four lanes in a pattern that you will record in your book. Close the covers , and run the gel for 2 hours at 55 watts (until the bromophenol blue dye has migrated to the end of the gel). After 2 hours , repeat the loading procedure in 4 new lanes. Run the gel further for 2 hours.

10.   When the gel has finished running , stop the run by turning off the power supply and disconnecting the leads. Drain the upper reservoir with a 60 cc syringe and remove the gel plates by releasing the clamps. Note that the lower buffer chamber will be radioactive because incorporated ^{35}S-dATP and short labeled products will migrate off the end of the gel. The lower reservoir is detachable to allow easy disposal of the radioactive buffer down the sink (in fumehood-flush with plenty of tap water).

11.   With the gel plate apparatus out, the plates will be carefully pried apart. Because , only one of the plates has been sigmacoted , the gel should preferentially stick to one of the plates. The thin acrylamide gel will then be carefully transferred to filter paper (this will be demonstrated first), and covered with saran-wrap.

12.   Gels will then be dried down using a gel drier, and subjected to autoradiography for visualization (we will dry our gels and put them to film)

13.   The data will be given to you in the next experiment class.

BUFFERS AND SOLUTIONS:

10X Electrophoresis Buffer

164g Trizma base.

27.5g boric acid.

7.45g Na_2EDTA

dH_2O to 1L.

CHAPTER 12

GENOMIC DNA PREPARATION.

Genomic DNA involves isolation, purification and precipitation. Genomic DNA? What is it? It is the whole shabang! You are trying to isolate all DNA from your cells. As a result, it tends to be pretty big(i.e. for human cells, the genome is composed of 23 pairs of chromosomes, with a total of 3.3billion base pairs– very big). Why are you getting genomic DNA? – A couple of reasons:

1. **O.J Simpson /Billy Clinton example**: This is the stuff you use to compare blood/semen samples. You isolate the genomic DNA ,

then perform some technique that can give you a reproducible and hopefully unique set of data (i.e. band profile on a southern for example).

2. **First attempt at cloning a new gene**: If you plan on looking for a completely new gene, you will have to have a genomic source of DNA material to look at. Although, you may be able to get protein or mRNA information before this step, you will ultimately need to define it within a genomic context (i.e. what chromosome is it on , promoter regions, etc, etc.)

3. **Checking genome for the presence of a particular piece of DNA**: Lots of examples in basic research, i.e. transgenic studies. You have made a transgenic/ knockout mouse. You want to check if, you actually put the gene in/or knocked it out. You need the DNA to look for it. In studying genetic diseases, you find a unique individual and you suspect the phenotype is caused by deficiency in such and such a gene.

4. **Studying genome for clues into gene expression and regulation**: Usually when people work with DNA , they actually work with a streamlined copy of their gene called the cDNA. This is a direct record of the messenger RNA responsible for coding the protein. Whilst , this is generally much easier to work with . it misses out a lot of stuff that may be interesting or relevant to your research ie. In the study of gene control, there is need to look at promoter /enhancer elements , transcription binding sequences etc. etc. You need genomic DNA in these studies.

NOTE: Genomic DNA ; what does it look like? Looks like snot.

Some basics includes the following points:

1. DNA is negatively charged.
2. DNA has lots of ring structures, therefore quite hydrophobic.
3. DNA is a robust molecule and is generally , the easiest biological macromolecule to work with.

NOTE:

1. Get it as pure as you can:

(i) If you have contamination (by proteins, or whatever) subsequent steps may be hindered. i.e. restriction digests, PCR, etc. etc.). You essentially don't want to introduce reasons for subsequent steps to go wrong through some mystery contaminant.

(ii) DNA are not like nucleases. Therefore, get it as pure as you can. Nucleases are everywhere. Be careful in handling the materials. Keep everything cold since these enzymes are much more active at physiological temperatures (i.e. use of "ice cold" this and that"). Generally , it is a good idea to wear gloves.

2. **Large pieces of DNA can shear:** This is not actually a big deal for most procedures but why take the risk? . In this case , dealing with a piece of DNA that is approximately 100kb in size, there should be no vortexing, no vigorous pipetting. Snip the end of your tip before use.

3. **Need to think about how much you need:** Small amounts are generally easier to work with , machines are smaller but take less time to operate , etc. etc. So, if you know , you don't need much , then only use that amount. In general, you will get about 200 g of DNA per 1×10^8 cells . Equivalent to about 100 mg of tissue.

GENERAL PROCEDURE: This is broken down to the following components:

1. **Get the cells ready:** Keep it cold and clean the cells up a bit. Our cells were grown in tissue culture media which is full of proteins and other nutrients. This extra stuff could be a problem if you do not purify it out. Therefore a good and relatively easy initial step is to rinse the cells out . This is what the PBS is doing.

2. **Lyse the cells open**: This is done to get at your DNA as it is inside the cell. It makes DNA easily accessible for subsequent steps. Lyse cells with Tris (buffering agent which is good at pH 6-8.   EDTA chelates divalent cations which are necessary cofactors for DNase activity (way of shutting down nucleases). NaCl at physiological concentration (generally considered to be 100- 150mM) . Keep all molecules happy (particularly proteinase K) which prevents unwanted aggregation.
SDS, nasty ionic detergent which is good at breaking membrane , general denaturant inhibit enzyme activity. Since

DNA is so robust , not really adversely affected by SDS treatment.

NOTE: When dealing with plant material, a very common detergent in place of SDS is Sarkosyl. This essentially behaves in a similar manner to SDS. Proteinase K , serine protease works well at 55 °C and is used because it is very effective and not particularly susceptible to SDS, and ether denaturants such as urea. Proteinase K will chew up protein , which helps lysis in general and frees up the DNA from any protein gunk associated with it. Best used FRESH. (*Quite an important step).

Incubation step generally a minimum of an hour . Most procedures go longer depending on the material being used (Most procedures outline an overnight or >16 hours inoculation time). i.e. mouse tails may want to go overnight. Lysis step is probably the most varied in literature regarding ways of extracting genomic DNA. It is important to realize there are many variations of the lysis procedure . Some are quicker , some are more efficient , some are more expensive , some only work in certain situations . The SDS/protein K is a very standard procedure.

CHAPTER 13

PURIFICATION OF DNA

Note that phenol/chloroform works with any nucleic acid preparation. O.K . Cell is lysed. Do you want to purify it now? **Phenol**

Chloroform Steps: ***Phenol*** is pretty nasty!. It will burn. (Apparently, someone once told me that if you cover 5% of your body, you will die) . Since we are using reasonable amounts, do Phenol, addition in fumehood. If you get it on you, not panic mode. Quickly rinse with off with cold water. Phenol is usually buffer saturated, i.e. when you buy it or make it , the solution comes in two layers, top layer is excess buffer, bottom layer is buffer saturated phenol. The pH of buffer important, acidic pH, DNA is soluble in phenol. If not, buffer saturated , any additional aqueous (i.e. water) solution you add will combine with phenol solution . This is not good because you want the layering effect. In order for your purification to work , we need the Phenol at neutral pH.

This purification procedure works on the principle of "differential mobility'. To your lysate, you will add phenol/chloroform /isoamyl alcohol usually at a volume ratio of 24: 23: 1. Phenol –organic solvent / nucleic acids are not soluble at all. Therefore, DNA/RNA will stay dissolved in aqueous phase. Proteins will also selectively go into phenol solution. Furthermore, phenol also acts a denaturant, proteins denature and form aggregates and will collect at the interphase. You will see GUNK @the interphase. Chloroform , also has same general attributes as phenol (as far as solvent properties) but also stabilizes the rather unstable boundary between aqueous and organic layers .

Isoamyl alcohol also contributes to interphase stability and also helps to prevent frothing. Generally , you do this step, 2 or 3 times .

The more times you do it, the cleaner your sample. You may even know that the interphase gets cleaner and cleaner with each step. Note this procedure is very reliable and does not lose much DNA yield. This is probably why a lot of laboratories still like to use it.

Sometimes, do a final chloroform step. Same idea. Interphase is a little trickier to handle. Be careful not to get any organic liquid in this last step, because it may carry over to the final product. I suspect that you finish with a chloroform step because it evaporates easily (?). In other words, this gives you the option of leaving the lid open to really make sure all your organics are gone from your preparation.

CHAPTER 14

PRECIPITATION OF DNA

This protocol will work with any nucleic acid preparation. Plus Ammonium Acetate; 0.5 volume. Why? It helps in the precipitation of DNA in EtOH. Can use NaCl, can omit entirely depending on the concentration of DNA. Salt will help neutralize negative charge of DNA which will also sequester the solvent molecules , in this case water. Use 100% EtOH. Time frame should be shown with graph. EtOH generally helps because it is much crappier solvent than something like water which is more polar.

Efficiency of EtOH precipitation is dependent on a number of things like temperature , time and amount of DNA. Mention isopropanol which people use for precipitation step, RNA tends to stay soluble in this solvent. Some people use it for this purpose. Use glass Pasteur pipette to spool DNA out carefully . Dip 70% ethanol, and resuspend in TE. The glass pipette technique seems to be favoured solely for SPEED . It is very quick to retrieve your DNA and also to wash it.

NOTE: The 70% ethanol wash is included to get rid of excess salts.

ALTERNATIVE: Some laboratories like to centrifuge the DNA pellet down , rather than spool it out. This may be more effective , if you have very low yield. However, you will still need to wash the pellet in 70% ethanol and respin your sample.

CHAPTER 15

NUCLEIC ACID QUANTITATION

Spectrophotometry readings: Using UV absorbance to ascertain DNA/RNA amounts and purity , ring structures can absorb UV wavelengths. Lots of things have ring structures including nucleic acids, proteins, organics, detergents, lipids, the list goes on and on. i.e. Both DNA and RN ring structure (See Merck) fluoresce strongly at – 260 nm and –280nm. For proteins, some amino acids like W, P and Y also fluoresce strongly at –260nm to –280nm.

Bottom line is that you have to take numbers with a grain of salt. Very rough estimate , since lots of things absorb at UV wavelengths. Also quite sensitive to pH . You can also assess purity by looking at ratios 260 over 280.

DNA A260/A280 = approx. 1.8

RNA A260/A280 approx. 2.0.

These are good ratios for purified product . If you get good numbers here, then , maybe you can also get good . Note that the conversion for O.D numbers to DNA amounts is the constant 50 ug /ml per O.D 250 mm value. Likewise, the constant for RNA is 40ug /ml per O.D . 250mm value. Genequant Pharmacia Machine sucks.

CHAPTER 16

RESTRICTION DIGEST

HAIKU- To help remember the importance of using and keeping these enzymes under -20 °C conditions. Restriction enzymes: Always work in cold weather. My haiku sucks. Splash.

Enzymes are pretty expensive so you have to keep them cold and keep them clean. This means that you always use a freezer box when keeping them at your bench. This means that if you don't have a freezer box, you do everything in the freezer. No double dipping.

If you will be using etBr, it is a carcinogen, so please be careful. Why are we doing it? We are doing it with our genomic DNA prep because we wish to cut up this huge piece of DNA so that we can have a population of DNA fragments of varying lengths. Essentially, we are hoping to make our data easier to interpret because these smaller pieces can be distinguishable from each other by size. Since our DNA is so big, we will literally cut it up very specifically into many, many , many fragments which will hopefully look like a smear on our gel.

Get yourself some catalogs from molecular biology companies, very handy to have. New England biolabs is especially good.

Restriction digests are probably the most common technique out there involving molecular biology . Chances are , if you were an undergraduate student , doing some project in the laboratory, you would have done some of these stuff. They are pretty straightforward but there are some useful nuances worth knowing about. This brings us back to the beauty of DNA work, you have lots of toys including restriction endonucleases.

Restriction digests; restriction enzymes or restriction endonucleases (RE) cut at specific recognition sequences and cut the DNA to leave " sticky" or "blunt " ends.

All started in 1970. Hamilton Smith at John Hopkins University studying *Hemophilus influenza* . Extracts from this bug could cleave DNA at very precise points turns out. Bugs have sets of restriction enzymes which are believed to serve as their "immune system" of sorts protecting them from virus infection (viral DNA gets cleaved). There are approximately 2500 different enzymes to choose from , which are found from screening more than more than 10,000 bacteria, eukaryotic restriction enzyme and alga chorella.

HOW TO USE RESTRICTION ENDONUCLEASES (RE)

Think about the following when using restriction endonucleases (RE).

1. **Know which one to use (most important aspect)**: This will largely depend upon what you want to do. We will go into this in more details when we go into the cloning section. For example, we are using a 6-cutter (recognizes a 6bp site) because this will cut the genomic DNA into a slurry of DNA fragments of all sorts of sizes. This is important for our southern blot because if we cut with an enzyme that cut more often, then we would be restricted to only small fragments. For the best enzymes or particular sizes of genomes , you can look up charts that give you information.

2. **Use very simple amount**: One unit of enzyme is enough (theoretically) to cleave 1ug of DNA in one hour at the defined temperature in a reaction volume <50 liters. But in reality, every enzyme comes with its own personality which can also be dependent on the batch or company they come from. In general, ask around, there are often tales of folklore that are around i.e. NEB Bam H1 sucks, use GIBCO stuff, like that.

Common restriction enzymes generally work quite well i.e. the ones that are always listed in vectors. As a rule, blunt cutters seem to be more finicky overall. Usually add more enzyme than needed. It is usually OK to do this and hopefully will ensure that the reaction will work even better. It is practically impossible to pipette anything less than 1.01 accurately in any way.

3. **Conditions of reaction**: Enzymes will always come with "special" buffer. What the buffer is called depends on the company. Generally buffers keep enzymes at a particular pH- around 7–8, a particular salt concentration , and has Mg in it. Things to make it happy. It comes 10x concentrated , which means you have to dilute it in 10 fold like when you buy concentrated orange juice and dilute it with water. Olden days buffers used to be separated singly as low salt, medium salt and high salt concentrations.

Salt is the most varied and arguably the most important consideration in getting a reaction to work. This is particularly true , if you wanted to cut your DNA with two different Res (double digest). In the past , you will cut first with RE that worked at low salt concentration / bring up the salt for second enzyme. This would be an example of a tough dilution question. Nowadays , there are handy dandy buffer activity charts that lists percentage activity for an enzyme under different buffers.

OTHER THINGS TO ADD IN THE PREPARATION

RNAseI: This gets rid of extra RNA in your preparation.

RSA: This is albumen or generic protein. Proteins have an uncanny ability to destabilize when solvated in a very dilute concentration. This is because there is too much water solvating protein therefore there is too much room to move about as proteins are very dynamic structures. This problem can be solved by adding more protein in the solution so that the overall quantity of protein is higher.

Spermidine: This is not used as much anymore. It is the eye of newt compound found in sperm. It is added because of its ability to interact with polymerases and other DNA binding proteins. What happens is that it frees up the DNA for RE.

DTT: Dithiothrietol (DTT) is a reducing agent. It is used during purification process by companies. They may get unwanted disulfide bond formation which would affect enzyme structure.

TIME AND TEMPERATURE GUIDELINES.

One hour is the bare minimum . It is considered safe to leave digests going overnight. It may even be better thing to do , if it is crucial for you to get complete digest. Some enzymes are sensitive to high and this is very useful , if you want to inactivate it.

4. **Your DNA sample**: There should be no organics , i.e. phenol or chloroform. They affect RE viability.

Methylation: This will also affect the ability to be cleaved. In actual fact, this is what protects the bacteria from its own restriction enzyme defence system, i.e. it has a corresponding methylation enzyme that protects bacterial DNA. This is why most *E.coli* strains that you work with have dam–and dem–nomenclature. This means that these methylation enzymes have been knocked out so that they won't affect the ability to cut DNA isolated from these bacteria. This is an important point since all organisms are capable of methylation to some degree. This is also the fallback excuse as to why something isn't working properly.

5. **Star activity:** This happens when an enzyme goes bad. Sometimes an enzyme will get sloppy and cut at sites that do not adhere to recognition sequence. This does actually happen with Bam H1 and EcoR1 and several others. Remember, these are enzymes and think of the sequence specificity as its substrate. Things that can cause non-specific cleaving include things that can affect the enzyme –substrate relationship.

6. **High glycerol:** Be careful when you add too much enzyme since enzyme is kept in a glycerol solution. Weigh too much enzyme >100 units/ug. Use too little salt < 25mM.

Note pH variations especially if too high >8 , presence of organic solvents and use of different cofactor i.e. Mn $^{++}$, Zn $^{++}$ instead of Mg.

7. **Stopping reaction**: EDTA: Heat the sucker . Some enzymes are heat sensitive . Stick in a 65 o C water bath for 10 mins. If you want to completely get rid of the enzyme , do a Phenol /CHCl extraction.

CHAPTER 17
GEL ELECTROPHORESIS.

Electrophoresis can be carried out using agarose, a polysaccharide polymer. It is used because of its ability to form pore sizes capable of resolving approximately 0.2kb to approximately 60kb. Check the table in this book. A big mesh of fibre is present and your DNA has to pass through it.

Electrical charge is the driving force and things will separate according to size. Lots of things to think about if you really want to impress people with shop talk but generally things like voltage buffer not that important in the grand scheme of things.

DNA works well because for things to separate in correlative manner, you do not need to worry about charge, density and shape. DNA only composed of four things of similar charge. Over 200– 60,000 bp, the charge density evens out very nice. Also, in our case, the DNA we are looking at has been cut with restriction enzymes. Therefore all DNA fragments are predominantly linear in shape.

BUFFERS: Loading buffer; Glycerol: thickens sample up so that it doesn't float away after you load it into the well. 0.1 M EDTA stop reagent for the assay. 1% SDS help denature the RE . Stop the reaction. 0.1% bromophenol blue. Dye helps you visualize sample when loading. It will run towards electrode. We can use it as a rough idea of where your DNA may be running. It is dependent on gel percent.

RUNNING BUFFER: Tris Borate EDTA → TBE. Borate provides ion which allows the generation of an electric field in the gel set up. TAE > Acetate as alternative ion is often used because it works and is much cheaper. Need 50 °C for 10l of TAE (Needs $10 for 10l of TBE). But tris/borate has a significant better buffering capacity , which means gel running is more reliable especially at high voltages , or long running times. But borate (when preparing the gel in microwave or oven) also forms complexes with the agarose sugar monomers / polymers. It can be a problem if using procedures to isolate band from a gel (low melt or use of chaotrophic salts).

VISUALIZING THE DNA: Use ethidium bromide (carcinogen), that interchelates DNA (Which has the added ability of uncoiling it) i.e. will affect the apparent weight very slightly (But not significantly) . Use UV light to see it. Some labs add EtBr into gel. Some add it after gel has run (i.e. stain with solution containing EtBr) . Adding it into the gel is much easier , but if the apparent molecular weight of closed circular DNA is particularly important to you, it may be worth adding after so that it doesn't affect its molecular weight.

ALL ABOUT SOUTHERNS AND NORTHERNS.

Southern and Northern procedures are all about putting your data on a membrane. WHY? Because a membrane is much easier to handle than the gel itself (stuff won't diffuse, won't rip easily etc. So, using membranes , you need to wear gloves . Two main types of membranes , nitrocellulose and nylon based membranes . Nylon is very easy to handle than nitrocellulose which can get quite brittle. Basic idea is to stick nucleic acid onto membrane . Membranes are hydrophobic and generally have a net negative charge (can get special membranes which are positively charged . This is actually what we use for our southerns. Charge does play a role and will dictate what type of incubations our agarose gel will need to get through before the transfer set up. However, interaction between blot and DNA/RNA is largely mediated by hydrophobic interactions (ring structures). The assessibility of ring structures is therefore important . RNA is capable of tertiary structure, and affects assessibility of ring structures. Dsdna alpha helix structure sucks as far as hydrophobic assessibility. That is why, in terms of ability to adhere to membrane ssDNA >>>RNA >>>dsDNA .(Should be demonstrated).

Baking membrane dehydrates sample. Molecules in solution are surrounded by a shell of solvent (in this case, water). This will hinder your hydrophobic interaction . Therefore, removal of all water will increase the hydrophobic interaction. Nucleic acid stays stuck better. Option: UV crosslinking. Nucleic acid covalent bond membrane. Particularly useful if you want to probe the same blot over and over again with different stuff. Nucleic acid will stay stuck.

MORE INFORMATION ON THE SOUTHERN BLOT.

HCl incubation and partial purination:Hydrolyses bonds between purine (rig structures of A and G) and the deoxyribose sugar. Normally, huge pieces of DNA get broken into smaller pieces which tend to adhere to the membrane much better.

NaOH: Disrupts hydrogen bonding which is the basis of double strand formation. Therefore DNA is now single stranded. Which binds better and is required if trying to probe DNA with a primer.

20XSSC: Use high salt NaCl sodium citrate (SSC stuff) . In the case of normal membranes, the membrane and DNA/ RNA is negatively charged. High salt negates this so that two can meet more efficiently. Sodium citrate, good buffering capabilities at high temperature which is what the hybridizations are done at. pH is altered by temperature and nucleic acids are sensitive pH especially RNA. We skipped the 20x SSC step in our southern protocol (i.e. the transfer was done in a NaOH solution). Idea is that we don't need high salt conditions since we were using a positively charged membrane.

WILL TRANSFERING BY CAPILLARY ACTION: Using good old diffusion to transfer material from one place to another.

SPECIAL SECTION ON REPLICATION. This is useful to know, since it will come up in probe production, sequencing and PCR techniques. Use magnetic board. You have your dsDNA (red elastic bands). Well aware that the ends are defined as 5 (phosphate group), and 3 (hydroxide group).

IMPORTANT POINT #1: All DNA Polymerase, add NTPs to the 3 end (therefore they work from 5 to 3 OR you can say they READ the template DNA from 3 to 5).

IMPORTANT POINT#2: All DNA polymerases need a primer . Need an existing piece of nucleic acid where it can add something to its 3 end.

So, replication is a concerted effort of a large number of different molecules. Polymerases makes the DNA. Primases : makes a primer (RNA) so that the polymerases can make the DNA. Helicases: opens the dsDNA up so that all this crap can get to the DNA.

Topoisomerases: Makes sure the DNA isn't structurally unhappy because of coiling stresses. First thing to happen is that the DNA needs to be opened up. Helicase does this (DnaB). BIG COMPLEX (Blown up rubber glove) , DNA polymerase III +helicase+ primase can get in. Primase (RNA polymerase) makes a short RNA segment (<5bp). Big enough for the pol III to work on. Helicase will keep DNA open as DNA polymerase travels down. REMEMBER: Must transcribe from 5 to 3.

ONE side. How it works is easy to visualize . Other side is trickier . (Backtrack a bit).

DNA pol III is a dimer. Therefore it transcribes the other DNA strand as well. But because it can only go a certain direction, it can only make small fragments at a time. Highlight leading vs lagging strand terminology.

Topoisomerases: show rubber hand analogy. Cuts and relegates to relieve stress from coiling. OK, if you leave it like this, you have two ds duplexes. One is predominantly dsDNA with a short piece of RNA at the beginning. The other has a strand which is made up of lots of smaller fragments that are half RNA half DNA.

This is where DNA polymerase 1 comes in (Barney doll). Special because it also has 5 to 3 exonuclease activity. (Use barney because he has teeth). So , it looks for an unligated region). Attaches onto the DNA preceding the RNA bit. And removes the RNA portion as it adds to the DNA portion. Therefore all DNA, then there is a ligation step (roll of tape).

PREPARING PROBES FOR SOUTHERNS AND NORTHERNS.

Making probes : Introduce the probes we are trying to make:

1. (for both Southern and Northern Blot) PGMR1 contain sequences of 28s ribosome from soybean (these molecules are very conserved).

FIRST: You will need a piece of DNA . This will be a piece of DNA containing the sequence that you are interested in exploring. Depending on what you are using it for, you may want to make sure , it is very pure. (the purer the probe, then the cleaner your results will be!).

We will be using a kit available from Amersham. It contains a solution labeled as Redprime and is based on the use of random population of hexa- nucleotides. In effect, this random population of oligonucleotides ensures that at least a small minority can be used as primers for DNA replication. The Rediprime solution also has a DNA polymerase (an altered form of DNA pol1) which does the replication. All you need to add are labeled nucleotides (this was our radioactive NTPs). After the synthesis of labeled DNA (Basically our probe), we next need a procedure that will allow us to separate the labeled probed from individual (radioactive) nucleotides. This was done using the Nick columns which work on the basis of gel filtration - we will cover that later.

An extra material, will also mention the NICK TRANSLATION METHOD. This method is a standard procedure for making a double stranded probe. This is worth mentioning because it is still commonly used . What is it all about. It is all about two DNA modifying enzymes (It's worth double checking the notes on replication-last week).

DNASE 1 AND DNA POLYMERASE 1 : (SHOW DNA pol O/H).

DNASE 1: Cuts DNA - does not specifically cut at 3 or 5 but rather will cut into middle of DNA . Will actually break the phosphodiester bond between two nucleotides in the middle of a strand of DNA.

SHOW O/H OF NICK TRANSLATION: You have a piece of DNA.

You nick it with Dnase 1. You bring in DNA pol 1 with labeled NTPs. DNA pol1 will chew up existing DNA and simultaneously add on labeled DNA.

COUPLE THINGS: DNA pol 1 is a shitty polymerase, will only latch on for about 10nt - this turns out to be perfect for this procedure. How well it works greatly depends on the amount of Dnase 1 added and time allowed to work. Too much nicking, fragments are too small, lose stringency. Generally , you want fragments to be from 400 bp to 1500 bp. Ie. DNA pol 1 w/o NTPs, the two exonuclease activities will predominate. Sometimes can buy anything called "KLENOW" fragment. This is DNA pol 1 without the 5-3 exonuclease activity . I will mention this enzyme later.

MORE ABOUT SOUTHERNS AND NORTHERNS: These are stuff on the whole southern and northern procedure. The fact that we have used a nylon membrane plays a big part in many of the steps of yesterdays and todays procedure. The bottom line is that nylon is really good for binding DNA (and protein). Therefore you need to ensure that your probe doesn't bind non-specifically. In the pre-hybridization step, your membrane is half empty. There are lots of areas where there isn't DNA on the membrane. Therefore these parts of a membrane are completely exposed . If you add the probe now , then the probe would generally have no problems binding to these exposed areas.

Therefore, you need to make sure that these areas are covered up. Use generic DNA to cover up these spots. People generally use salmon sperm genomic DNA. You basically want something from unrelated species. There are lower chances of cross-reactivity but what about conserved sequences like 28s ribosomal DNA?. The DNA is also usually sheared to the point that it loses any specificity . Shearing is generally accomplished by vigorous pipetting using a large guage needle .

CONTENTS OF THE VARIOUS PREHYBRIDIZATION AND HYBRIDIZATION SOLUTIONS.

Formaldehyde: This is a compound that is used extensively by some laboratories when using DNA for blotting. Apparently useful because it displaces water from the oligo nucleotide structures. It makes nucleic acid more rigid. Therefore, your DNA or RNA strands are more like to be open than curled up in a ball. Therefore, it is less likely for the occurrence of non-specific binding due to clumping and aggregate effects. There is better stringency.

Denhardt's solution: Think of this as a streamlined blocking reagent. It has BSA, Ficoll (big sugar) and polyvinylpyrrolidone (even bigger polymer).

Dextran sulfate: You can use this or polyethylene glycol. It sucks up water. It artificially increases your probe concentration by sequestering water. Analogous to adding more probe, 10% should increase rate by ten fold.

SDS: This is a detergent. It alleviates hydrophobic interactions resulting in non-specific interactions. Remember, two pieces of nucleic acid can stick together simply by virtue of their hydrophobic nature.

OTHER THINGS TO THINK ABOUT IN THE HYBRIDIZATION STEP.

The probe: This is all about kinetics of association. The more you add, binding will be quicker. There is lower specificity. Probe details include:

1. Labeling: The amount of probe will also depend on how labeled your probe is. You are going to see it. It varies from system to system. Using a biotin ATP system, you need approximately 20 to 50 ng probe/ml of hybridization solution. For freshly labeled radioactive probes, you probably need approximately 1-5 ng/ml. Consequently, radioactive probes are much more sensitive , but are generally need to be used within 2 days of production. Chemically modified probes like the biotin system or DIG system, can be kept and stored for much periods of time.

2. Length: The lower limit is about 20bp long. The longer, the more specific , unless of course , you are looking for homologous sequence that may not be identical.

The Stringency: You need to get comfortable with this concept. High stringency means that the incubation is done in too strong a setting, such that nothing (including the probe) will interact with its complementary sequence. Low stringency means that the incubations are too weak and your probe wants to bind to everything. Stringency is the difference between perfect data , dirty data or no data. Couple of things that affect stringency are (Na); the salt concentration, (Formaldehyde) all relate to specificity of probe and Temperature of incubation. The lower the salt, the more stringent the condition. The higher the percentage formaldehyde, the more stringent the condition. The higher the temperature, the more stringent the condition.

Most laboratories with their own version of "stringency" , have conditions that are classed as high, medium, or low stringency. For example; High would be something like 42 °C with 50% formamide or 68 °C no formamide. Low might be 30 °C no formamide.

High frequency conditions are usually considered appropriate for blots when a long probe greater than 200 bp is used to detect a specific piece of nucleic acid (ie. Vast majority of nucleotides are complementary. But sometimes, you may want to play around with your stringency values i.e . using degenerate probes, or short oligos as a probe. The easiest thing to alter is your temperature.

You can use the following as guideline for temperature:

1. Figure out the melting temperature of your duplex- probe target nucleic acid interaction. Then you would want to use a hybridization temperature that is approximately 12 to 20 ° C less (no formamide). But to do this, you may have to resort to using any number of scary equations that are out there in the literature. Generally not worth the effort, unless you resort to an internet site to help out. The bottom line is that stringency conditions will vary from situation to situation. But hopefully, the above will give you some guideline to pursue. Follow these guidelines, make an educated guess, look at your data and adjust the conditions accordingly.

2. **The time of hybridization:** Generally not mucked about much. Usually people go O/N. Some argue that under long incubation times, the two strands of our probe may reanneal such that there is no probe left to bind to DNA on the membrane itself.

3. **The Washes:** After hybridization, you have done a series of washes that are generally designed to make sure excess probe is gone from your membrane. You will notice that it

has high detergent (SDS) and low salt (0.1 X SSC). Now our blot is ready for development using autorad film.

If however, you are using a non-radioactive system (i.e. Biotin system) , then there is an added BSA block step. This extra step is needed here because the detection system involves the addition of strept- avidin complexed alkaline phosphatase. This is because, membrane is nylon. Nylon also loves protein . Therefore, it is a precautionary measure. We add an incubation step where the membrane is additionally coated with a generic protein. The fact that these non-radioactive systems have an extra step requiring protein interactions , is a big factor in why some laboratories say they result in high background. This is an alternate system to the Biotin system.

The chemilluminescence autorad film , the one we are using, incubate reaction for 1 hour. (Better than 3 to 5 hours). Then expose the film . Good for approximately 20 hours after DIG system. For Digoxigenin labeled NTPs, use an antibody conjugated with AP. This is alternate to the biotin system. Reaction time is about 10 min. Then expose the film. Good for days.

4. **Stripping blots:** Incubate the blot at high temperature, low (Na) i.e. 80 o C – 90 o C, 0.1% SDS, 0.1 x SSC, 65 ° C.

5. **Vectors:** This is a very general heading of "cloning"– moving DNA around. Many types of vectors have a couple of common traits.

6. **Upkeep:** This can replicate independently of integration into host genome. You can use existing replication machinery of host organism or you use a special organism that has been previously genetically engineered to give you the replication of transcriptive machinery.

7. **Place to put DNA in:** All vectors generally have defined places where you can insert DNA into. This DNA is of course, the type of stuff that you are interested in. i.e. a piece of your gene, a chunk of chromosome where you think something you are looking for is in.

8. **Something that lets you know if the Vector is in the organism:** All vectors will generally contain a selective marker. Usually resistant to some sort of drugs . Common ones used in research field are ampicillin and neomycin. Some organisms will rely on auxotrophic traits. i.e. Yeast that is mutant so that it can't make up. Therefore you need to supply it in media. Well, yeast expression vectors will have a gene that allows these auxotrophic yeast to make their own up.

9. **Something that lets you know if the DNA you put is actually there:** Some strategy that lets you know if your DNA of interest actually made it into the vector is that usually some form of insertional inactivation , or size constraints i.e. with bacteriophage methods. If no DNA gets in , the lamdba vector is too small to functionally replicate and be packaged. You need to know which vector to use, your first

constraints and how big a piece of DNA you will need. Depending on the size of the piece of DNA , you are interested in, then the type of vector you use is important.

SPECIFIC TYPES OF VECTORS

1. **Plasmid vectors:** These are vectors such as Puc18. There are couple of things you should know:

   a. Selected marker: This involves use of drug resistance drugs such as ampicillin. Therefore, if you want to ensure that your bug has gotten the plasmid , then you just test for the bug's ability to survive under drug treatment.

   b. Blue white screening: This is a little more sophisticated. Based on the following idea, things to point out are LacZ gene (b-galactosidase). Bugs will constitutively express a lac repressor which interacts with lacUV promoter. Therefore, no transcription of beta galactosidase , but if you add lactose, lac repressors fall off the promoter. What you usually use is IPTG (Isopropyl -b- D-thiogalactoside) which is an analog of lactose.

   Anyway, if there is no insert, lacZ gene will be transcribed. It codes for b-galactosidase. If there is an insert , something else will get made. Frame shift mutations will not get lac Z to be transcribed. Use of lac Z is called alpha complementation.

   c. Good opportunity to bring up multiple cloning site. Small area of the plasmid where the vector designers have decided to put lots of unique restriction sites in a small area. i.e. try to make it easy as possible for you to have a convenient nice restriction site that works for your cloning strategy.

EXPRESSION VECTORS: If you put your target DNA downstream of a good promoter capable of transcription , then you have an expression system. For example: 1"7 promoter system show OH. Prepare hand out of different vectors. Fusion protein is up to 50%.

2. **Bacteriophage vectors:** This is a fancy word for virus that infects bacteria. It is good in moving approximately 20 kb worth of DNA. Good example of bacteriophage vector is the "lambda DNA". This is about 49kb in size. Highlight essential verses non-essential regions of the bacteriophage genome. Essentially, the key idea is to take advantage of the viral pathways to get your DNA inside i.e . you can package your DNA into a bacteriophage structure which can infect and shunt your DNA into the host cell (a bug). Another way to look at this is to think of it as a more efficient means of transformation.

Once the DNA is inside, the bacteriophage will still go about making more DNA , and making all the necessary components to make more of itself (actually called the lytic cycle). This is how you get more of your DNA. The only technical difficulty is how you would "harvest material" i.e. your DNA is not in the bacteria per se , but rather is inside the bacteriophage , and the logistics of getting bacteriophage cultures are more difficult (i.e. plaques on a plate). But DNA isolation techniques from bacteriophages are very easy and proficient , resulting in better quality DNA preparations.

3. **Cosmid vectors:** Balderdash OH. Good for packaging 30– 45 kb of material. Think of it as an unworthy bacteriophage i.e. even essential regions have been deleted and this is why it can carry more than normal bacteriophage vectors. But specially cohesive sites (COS sites) are still present . Therefore, still has enough genetic information to get packaged but not enough to "be bacteriophage" i.e. doesn't have information to make more bacteriophage . Now you can use a kit t6o supply materials (bacteriophage bits and packing machinery) needed for packaging of your DNA into a "pseudo" bacteriophage. Then your bacteriophage can infect a host organism , and effectively transform it. (deliver your DNA). Once, DNA gets in , the COS sites will direct re- circularization and now you essentially have a Big Plasmid(therefore can use normal plasmid techniques for isolation).

4. **YACs and BACs:** The realization that the components of a eukaryotic chromosome that are required for stable replication and replication in yeast, are very small, very defined sequences. They are recombinant because they can make artificial chromosomes with is what the AC in YAC stands for.
It can take up to 1Mb. Otherwise, this sort of vectors has similar things. Usually yeast systems use auxotrophic selection process. Note that BACs are more common nowadays. Similar

to YACs is that they can take up a huge amount of DNA (up to 300kb). The difference is that you can work in *E.coli*, which is more easier than yeast manipulation.

WHAT ABOUT EUKARYOTES? Life is complicated. If you are interested in studying the effect of protein expression in the eukaryotic cell, it is not good enough to do this in bacteria because bugs can't do many post-translational modifications that are normally found in eukaryotic settings (i.e. some lipid modifications, glycosylation). Generally, the same kind of things apply here as well. Minor differences, none conceptually i.e . different resistance selection markers. Common ones are in neomycin and paramycin.

SHUTTLE VECTORS: This is designed to work in both *E. coli* and in eukaryotic cells. Therefore, all manipulations and mucking about happens in *E.coli* (because it is easy). When plasmid is ready, put it into eukaryotic cell for expression.

LIGATION REACTIONS: **Ligase**: Stick two pieces of compatible ends of DNA together. 5" end must be phosphorylated. Also, the ends that you are trying to put together must be compatible, i.e. if you are dealing with sticky ends (i.e. they have the overhangs.

The overhangs , must be complementary to each other. Blunt ends are special in that all blunt ends are compatible with each other. Ligation works much better for sticky ends than for blunt ends. Ligation procedure is usually done O/N at 16 °C.

There are some variations depending on the ends you are putting together.

We cannot directly check if ligation worked except by virtue of getting bugs to grow after transformation. It is useful to try different insert vector ratios. Molar ratios that are often tried are; 1: 1, 3: 1, 5: 1 etc. Generally, more insert is better, but every ligation is different.

There are equations you can use to figure out the molar equivalents of your DNA ends, to get good ligation numbers. It is kind of complicated and not many people use them.

CALF INTESTINAL PHOSPHATASE (CIP): Can also use alkaline phosphatase: Dephosphorylate 3 phosphates. This is useful to dephosphorylate cut vector. Essentially, a handy trick to prevent re-circularization of vector without insert (i.e. it is a way to make sure that any colonies you see on the ligation plates must have an insert in the multiple cloning site).

(The act of cutting out fragments and placing them into vectors for a particular purpose , is usually quite difficult to explain. It is like one big word puzzle , where you are trying to fit pieces together so that everything is compatible , in the right orientation, will get expressed properly , and will actually result in the product you care about. We will attempt to go through a cloning strategy in one of the other chapters.).

TRANSFORMATION: The story so far is that you have just done a ligation reaction where you have added a cut vector and a fragment together. You want to know if the ligation worked, first, you are going to put the ligation mix into bacteria , and plate the bacteria onto ampicillin supplemented media.

Therefore, in order for bugs to grow, it must have ampR. Therefore must have plasmid but plasmid must be circularized in order for it to get continuously and independently replicated, therefore, ligation has to work to get all of this. If the insert is in, use the white blue system . (Your plates will also have X-GAL). Transformation, transduction and transfection are all about getting DNA into an organism. Transformation is putting DNA into a prokaryote. Transduction is the use of infection (i.e. viruses) to get DNA into a eukaryote. With bugs, two main techniques, $CaCl_2$ and heat shock treatment and electroporation.

In both cases, you need to make competent cells. Cells that are primed and ready for the acceptance of DNA. Generally, involves a series of growth steps so that bugs are at just the right stage of growth.

In $CaCl_2$ method, competent cells are treated with $CaCl_2$. Essentially, the membranes are thrusted around . The bottom line is that cells are delicate. Ice step is believed to allow the DNA to adhere to membrane.

Heat shock may make the membrane move around more (fluid membrane). Membrane gets weaker, holes gets bigger. DNA falls in. Using 37 °C incubation allows thrashed cells to recuperate. It also gives time for plasmid to replicate so that bugs are ampR.

ELECTROPORATION: This will give much higher transformation efficiencies. ZAP is a current through the bugs. Again nobody really knows what is going on. A bit more versatile (i.e. most organisms do not have a heat shock procedure).

Microbiologists working on other bugs , pseudomonads, bacilli , Streptococcus e.t.c. e.t.c. All of this will translate to some magic current value which is what causes this whole thing to work. Because of this, you need to be careful with salt content in your cells or ligation mix. Making competent cells usually involves successive water washes. O/H,the only physics equation that will be shown in this lecture is $V=IR$.

If you are not careful because of salt, or air pockets, etc. , etc., You get something called Arcing. This is a small explosive happening , which usually consists of a bang (to varying degrees) cracking of the cuvette and a flash of light.

General scariness but not all dangerous.

To my knowledge, I have never seen an obituary which described cause of death by electroporation arcing.

TRANSDUCTION: I mentioned transduction before (i.e. with a bacteriophage).

TRANSFECTION: This is the act of getting DNA into eukaryotic cells. Also use electroporation. ZAP the buggers! If you still do not know how it works, use gene gun ballistic approach.

PURIFICATION KITS:

An example is gene clean. Nucleic acid purification kits usually involve the use of silica based beads which are specifically designed to interact with the very electrostatically charged nucleic acid molecules. In particular, our gene clean preparation is a good example of how kits work in general.

Our DNA sample was first treated with NaL. This is classed as a chaotrophic salt which is really a fancy way of labeling a salt capable of altering structures by interacting with many molecules of water. You need to envision your NaL as binding to several molecules of water at high stoichiometry. This loss of water from the DNA structure in particular will alter its shape , charge etc. and this specifically binds our beads (the white stuff).

We next do a wash, which in our case is called new wash buffer. We are not entirely sure what this is , but you can bet that it doesn't have much (if any) water. You don't want water during the wash steps, or else your DNA will get the water back and fall off the beads! This is why the final elusion steps are water alone. There are lots of variations of these kits. Popular these days are versions that are set up in a column format.

CHAPTER 18
PLASMID DNA ISOLATION

Isolating plasmid DNA from other types of DNA (i.e genomic) is actually very simple. In short, it usually involves a denaturing step, followed by a quick renaturing step. The idea is that plasmid DNA being much smaller, can renature relatively easily – consequently , once back to normal , it can go into solution easily. Something like genomic DNA will have an incredible hard time renaturing because it is simply too big and too complicated.

It doesn't renature effectively and instead tangles up and precipitates out. If you think about it, you have now separated your plasmid (in solution) from your genomic (out of solution) prep. You simply have to centrifuge away the genomic pellet , and you are left with your plasmid DNA (plus all other cellular crap like proteins etc. , etc., however , now you can use any standard DNA procedure).

ALTERNATIVE PLASMID PREPARATION METHOD: Alkaline lysis method is one of the most common ways of isolating plasmid DNA. When you get the DNA, you need to open the cells up using a treatment of *E. coli* with EDTA and sugar. Peptidoglycan layer of bacterial membranes requires Mg^{2+} for stable conformation. This is also what lysozyme directly attacks. Glucose is for osmotic process. Solution II NaOH and SDS does most of the work, ruptures cells, and denatures everything. Low pH specifically breaks Hydrogen (H^+) bonds in dsDNA. Now your test tube will now be this messy mix of denatured stuff: Genomic DNA (Big)= denatured; Plasmid DNA (Small) =denatured, and Proteins= denatured.

Throw in salt that is acidic (KAc pH4.8). Salt helps in the precipitation process. Acid causes things to go back to neutral

. DNA can renature but happens very quickly. Large DNA denatures as a mess (kind of like of Christmas lights) . Small DNA renatures, OK. So genomic DNA will precipitate out (should see a white mess), but your plasmid DNA will now be in solution. If you still want to clean up your prep, remember the phenol chloroform procedure. You can do it.

PROTOCOL FOR DNA EXTRACTION USING PHENOL (WHOLE BLOOD/ TISSUE) :

1. To 200ul of blood or a tissue pellet (in a 1.5ml of eppendorf tube). Add to the tube ; the same volume of extraction buffer (200 ul); 0.1 M Sodium chloride, 0.05M tris– HCL , pH 7.5, and 1Mm EDTA , 10% SDS. Proteinase K to a final concentration of 60 ug/ml, (0.6 mg/extraction buffer of 100ml).

2. Mix well and incubate at 37 ℃ overnight or at – 52 ℃ for 2 hours. Mix occasionally during the incubation to keep the tissue suspended.

3. Add an equal volume of phenol, pH 8.0 (to equal volume of each sample, add equal volume of phenol).

4. Mix gently for 10 minutes.

5. Centrifuge for 10 minutes at 13,000 rpm (14,000 g) in a micro centrifuge.

6. Carefully remove the aqueous layer to another tube and add half of the volume of phenol and half of chloroform – isoamyl alcohol for 10 minutes at 13,000 rpm (14,000 g) at 4 ℃.

7. Dry the pellet and re-suspend it in TE (25ul). This is the stock DNA. For PCR, dilute in 1: 10 ratio (90 ul of TE-10ul DNA template). Check the purity and quantity of the template DNA in a spectrometer. (already prepared in the ratio of 24) (i.e. sample 250 ul ; phenol 125ul, chloroform ; n-isoamyl alcohol 125 ul).

8. Repeat steps 4 and 5.

9. Carefully remove the aqueous layer to another tube, add 1/10 the same volume of 3M sodium acetate , pH 5.2 (one tenth of sample volume) and 2.5 times the sample volume of ice-cold 100% ethanol.

10. Precipitate the DNA 20 °C overnight.

11. Centrifuge for 10 minutes at 13,000 rpm (14,000 g) at 4 °C.) .

12. Wash the pellet twice with 200ul of 70% ice- cold ethanol, for each wash, centrifuge the pellet.

PLASMIDS CURING EXPERIMENT:

All isolate that exhibited ability to degrade cellulose were subjected to plasmids curing experiments. This was to enable one determine whether this cellulose degrading ability exhibited was chromosomal mediated. Curing was carried out by the method of Miller (1972) . The isolates were grown for 24 hours at 37 °C separately in nutrient broth, pH 7.5, in the presence and absence of 75 ug/ml acridine orange. Broth cultures were subcultured into plates of Nutrient agar , which were incubated at 37° C for 24 hours. Colonies from the agar plates were then screened for their ability to degrade cellulose as previously described.

All isolates that loose their cellulose degradation potentials after treatment with acridine orange was taken as evidence of plasmid mediated.

GENE TRANSFER BY MATING:

The method of Benneth and Richmond 1975 and Avila *et. al.* 1984 were used in the transfer. Cultures of donor and recipient bacteria were grown on a liquid medium of nutrient broth for 24 hours at 30 ºC. 1ml of each of the donor and recipient cells were mixed in 1ml of nutrient broth and the mixture incubated at 30 ºC for 180 minutes. Dilutions of the mixture were used to screen for cellulolytic ability using whatman #1 filter paper and the mineral salts. The experiment was set up for 4 days with both negative and positive controls.

The recipients used for mating were *Pseudomonas aeruginosa* and *Pseudomonas fluorescens.*

DETERMINATION OF FREQUENCY OF TRANSFER.

Colonies in the test plates (A) and control plates (B) were counted , the frequency of transfer of transfer of plasmid from donor to recipient. (F).

F= B/A X 100 Where A= Colonies on test plates. B= Colonies on control plates.

ISOLATION OF PLASMID DNA.

The method of Bennete *et. al*., 1986 was used. Bacteria from 1.5 ml of an overnight culture in nutrient broth were harvested by centrifugation in a 1.8ml polypropylene centrifugal tube (Eppendorf type) in a micro centrifuge. The cell pellet was resuspended in 100ul of lysis buffer (lysozyme + solution 1) . The enzyme was added to the buffer immediately before use. The suspension was left at room temperature for 15 minutes to allow weakening of the cell walls.

Lysis of the cells was accomplished by addition of 200ul of alkaline SDS (Solution 2) and mixing by inversion two or three times. The solution at this stage was clear and viscous. The large chromosomal DNA molecules were fragmented and denatured in alkaline solution.

After 5 minutes, the lysates were neutralized by the addition of 150ul of 3M sodium acetate, pH 4.8 (Solution 3) and mixed by inversion. The sodium acetate causes precipitation of the SDS and the denatured chromosomal DNA, which was then removed by centrifugation in a micro- centrifuge for 5 minutes. (12, 800g).

400ul of the clear supernatant fluid were removed to a clean Eppendorf tube. Plasmid DNA was precipitated by the addition of 1ml of ethanol chilled to -20 °C. To complete the precipitation of the DNA, the mixtures were immersed in liquid N_2 until they become just viscous . The DNA precipitates were collected by centrifugation for 20 minutes.

The orientation of the tube within the micro centrifuge was such that the feathery , translucent pellets were not disturbed when the supernatant fluid was removed by aspiration. The pellets (usually deposited up the side of the tube) were suspended in 400ml of Tris acetate (solution 4) and proteins were removed by extraction with an equal volume of trisaturated phenol.

The two phases were mixed by vortexing the tube briefly and were separated by centrifugation (2 minutes, 12,800g). The upper aqueous phase was taken into a clean tube , with care top avoid taking phenol over , and the DNA was re-precipitated by addition of 1ml ethanol , chilling to -20 °C.

Cooling in liquid N_2 and harvested by centrifugation. The supernatant fluid was removed by aspiration and the pellet washed with 200ul of diethyl ether (without mixing). After a 15 seconds centrifugation , the bulk of the ether was removed by aspiration. Residual ether was allowed to evaporate before the pellet was dissolved in 20ul of sterile distilled water.

The DNA solution was then mixed with a loading buffer. The samples were then ready for agarose gel electrophoresis.

CHARACTERISATION OF PLASMID DNA BY AGAROSE GEL ELECTROPHORESIS.

PREPARATION OF GEL AND GEL SLOTS: 0.82g of agarose was dissolved in 100mls of TEB buffer (40 mM Tris . HCL, 1mH EDTA, 50 mM Boric acid) by heating at 100 ° C for about 30 minutes on a magnetic stirrer until the agarose was completely dissolved and became clear. This was allowed to cool to about 60 $^\circ$ C and then carefully poured on top of the glass slab with the "comb" in place.

The well- set agarose on glass slab with the comb was transferred to the electrophoresis chamber and flooded with TEB Buffer.

The comb was then carefully removed creating 12 gel slots . 22ul of the DNA extract was loaded onto gel slots. A mixture of standard plasmids of known molecular weight was loaded onto one of the slots as control.

ELECTROPHORESIS OF PLASMIDS:

The power pack was connected to the tank and the electrophoresis allowed to proceed at a constant voltage (50 mv) until the dye reached bottom of the gel. At the end of the run, the gel was stained for 20 minutes in TEB containing 1mg/l ethidium bromide. DNA bands were visualized by illuminating the gel with an ultraviolet transilluminator. Photographs were taken of gels positioned over the UV light source. Polaroid type 55 films exposed through a Tiffen 15 orange filter was used.

The relative distance moved by the test plasmid DNA band was compared with that of standard plasmid run side by side. A graph of log of molecular weight of the known plasmid DNA against distance moved in millimeter was plotted.

The molecular weight of the unknown plasmid DNAs was then calculated using the known distances from the standard graph. The distance moved by plasmid DNA is inversely proportional to the log of the molecular weights.

In the electrophoresis of plasmids, the tracking dye will almost reach the anode end of the gel. The gel slab was removed into a solution of ethidium bromide (5ug/ml) in distilled water to stain. Staining was for one hour after which the gels were viewed under ultra-violet lamp to visualize the plasmid bands. The distance migrated by each plasmid was also measured. The gels were then photographed with a yellow filter.

To determine the molecular weights of plasmids, standard markers were also extracted and run along side the plasmids. The electrophoretic mobility of these plasmids were obtained and used to plot a calibration curve of logarithm of molecular weights against the electrophoretic mobility. This curve was then used to determine the sizes of plasmids by simple extrapolation.

CHAPTER 19

CURING OF PLASMIDS BY TREATMENT WITH SODIUM LAURYL SULPHATE (SDS).

The method of Tomoeda *et. al* (1968) was used . To 90ml of nutrient broth , 10g SDS was added. The suspension was autoclaved

, adjusted to pH 7.6, steamed for one hour and then preserved as a stock solution.

Overnight cultures in complete broth were diluted and 0.5ml volumes were added to nutrient broth of pH 7.6 and incubated . Two volumes of 10 % SDS stock solution was added to give the required final volume for the experiment.

CHAPTER 20

SEQUENCING PROTOCOL

Sequencing is a big part of genetics nowadays. You always hear about it in genome projects. How does it work? Saager method often called chain termination . The general gist is as follows: It shows O/H (go over general procedure which is also highlighted in your laboratory manual).

There are a couple of key components: They are as follows: Template DNA, Primer, DNA polymerase , Label , dNTPs and ddNTPs and electrophoresis system.

We will primarily focus on the template and primer that you use for DNA sequencing. The reality is that many people do their own sequencing these days because it is so cheap to send your stuff in for automated sequencing services . However, even if you do this, you will probably be preparing your own primer and your own template.

1. What is a template? This is the piece of DNA that you want to sequence.
   (a) 5'...3' sense template.
   (b) 3' ....5' sense template.

Let us assume that you want (b) .

So for a while, especially in the past, it was a rule that you needed a ssDNA template (in reality, it doesn't really matter which side you use). To get a single stranded template, you can use the phage vector M13. What is M13? It is a type of bacteriophage similar to lambda. Why is it special? Well as a phage, its nucleic acid exists in a ssDNA form . But whilst in the bug (i.e. *E. coli*), it exists as a dsDNA replicative form. This means, you can mess around with it (i.e. clone in a sequence of interest) like a normal plasmid while in its RF. But , you can then get single stranded product upon phage formation.

i.   Cut out plaques, diffuse out phage from the agar,and isolate ssDNA. (very clean).

ii.   Clean is good. However , remember that isolating a phage is a pain on the butt.

Other ways to get single stranded template include:

1. One enzymatic digestion of double stranded plasmid , and.

2. **Two asymmetric PCR**.

2. Ok. If you now have template , you now want a primer that can anneal with this single stranded template.

(a) 5'...3' sense template.

(b) 3' ....5' sense template.

You want the primer to be approximately 18 to 24 nucleotides long. And you want to bind specifically. Need to figure out melting temperature, then apply a temperature that is at least 20 °C in excess of the melting temperature.

−2(A + T) +4 (G +C) rough number (usually good enough). In our case , the Tm is approximately 45 °C . That is why we are using a beaker at 65 °C. Often, there are common primers (sometimes called universal primers) depending on the vector used for your template production. For example, the primer we are using is upstream of the MCS where the DNA of interest is put in.

3. NTPs and labeling P32 strong but bad resolution / S35 weaker but good resolution/ fluorescent dyes, dideoxynucleotides (ddNTPs) forced termination , are missing the 3" hydroxide group, which means that another NTP cannot be added on (Hence the name of termination of replication).

4. **The polymerase being used** : Many variations which are interchangeable are usually used on the "basis of what is available" . We are using something called sequenase which is a modified T 7 DNA polymerase where its inherent 3- 5 exonuclease activity has been removed. Sequenase is also

blessed with excellent , processivity , elongation , good incorporation , < 55 °C show O/H of many types of polymerases with various attributes. (i.e. does the polymerase have the following activities or characteristics?).

- **3 > 5 Exonuclease.**
- **5 > 3 Exonuclease.**
- **Processivity.**
- **Elongation rate.**
- **Incorporation of nucleotide analogs, dNTP, 7– deaza-dGTP.**
- **Temperature range.**

Use of thermal stable polymerases allows something called cycle sequencing. The main limitation of the method we are currently using (not cycle sequencing) is that , you are limited by the amount of template you have. For instance, if you have only 3 template strands, you will get 3 terminated products. Therefore to get enough different fragments at high enough intensities for detection, you need a lot of template.

An alternative is "cycle sequencing" which do reaction once. Then you can heat to denature labeled product from template, plus primer and do it again. Repeat until reagents are exhausted. The problem is that you need to think a bit more about labeling procedure. You cannot have separate labeling step, unless you want to add fresh reagents with each cycle. (bit of hassle). But if you just include a stock of labeled NTP in your mix, then you will get stronger signals for bigger bands. Therefore, you can either make the primer labeled, or you need to label the dideoxy-nucleotide . Therefore, labeling is uniform per fragment.

The other bonus about using heat stable polymerase like Taq or Vent, is that you can use double stranded template because heat will denature it into single strand form anyways.

COUPLE OF KEY POINTS RE-EMPHASIZED:

1. Template and primer must be very, very, very, clean. This is why, if you have a lot of problems , messing around with M13 to get ssDNA template may be more reliable.

2. Amounts, amounts, amounts of template/primer/ NTPs/labeled NTPs/ddNTPs are crucial , and will ultimately depend on the polymerase and labeling system you use. You usually don't think about this because manual sequencing is usually performed using " kits". But if things are working and you need "tweaking" of the system . It is good to do your homework , i.e. Taq polymerase does not like ddNTPs, so generally , the ddNTPs, so generally , the ddNTP/Dntp ratio is much higher , but check your technical info, because there are also special genetically engineered taq polymerases that are much more accommodating.

3. Primer- template annealing temperature is very important because if your primer binds in an area of non-consensus , you will get replication from different start sites which will

ultimately lead to different size fragments. Use your Tm= 2 (A+T) + 4(G+ C).

5.**Gel electrophoresis** : Three times to work rule is in effect here in a big way. This will probably be the most technically challenging thing , you need to do all year. Gel has to be perfect. No dust. Easy to come off, No bubbles. Detergent must be gone, but it will affect the polymerization rates.

Uneven polymerization, uneven looking gels , siliconize one of the gel plates, so that gel can come off the glass. Gel is composed of acrylamide (similar idea to agarose), mesh is much finer, resolution is much higher. Bis -acrylamide , temed/ ammonium persulfate and gel is prewarmed so that the mesh is consistent throughout. Sample is usually loaded in "stages". Draw the picture.

OTHER METHODS OF SEQUENCING:

Maxam and Gilbert: Go over real quick real . Go over overhead.

Xerox Enlargement Microscopy Technique: Ha, ha, ha. Very lame.

PROBLEMS: There is almost always problems because of the quality of your template because everything else is so standardized (kits, oligonucleotide production). If your template is dirty, it will affect how your reaction works. That is why some people suggest you use ssDNA template from phage , because nucleic acid isolation from phage is a very clean process (although nowadays , there are many good kits out there that can give you amazing pure DNA samples).

The sequence is finicky: high G+ C content (i.e. Tuberculosis) . GC compression is a big problem. Why because G- C interaction is very tight. Can cause problems. DNA has weird secondary structure which affects sequenase steps and how it will run on gel. Therefore, can cause problems with template and with products. Fix it with use of formamide in gel, higher temperatures during reactions and use of GTP analogs dNTP/7-DEAZA-dGTP.

AUTOMATED SEQUENCING:

O/H use of big dyes. ddNTPs are fluorescently labeled Taq / sequenase ; stress cleanliness of template and primer DNA ; show prices cheap , cheap! May need to wait for weeks for data. Can get approximately 500 bp of good sequencing data and approximately 300bp of data with manual method.

The best advantage of manual method is customized tweaking for tricky sequencies and you get the data in 2 days instead of 1-3 weeks. Show O/H of data.

STRATEGIES TO SEQUENCE BIGGER SEGMENTS:

Nest deletions: Pst 1-3 overhang (exonuclease III), show gel picture.

Hail Mary/Shotgun approach: Genome projects sharing.

CHAPTER 21

PROTEIN ANALYSIS TECHNIQUES

Protein work is the next level of complexity when comparing work done with DNA. Basic idea is that with DNA, you have got 4 different components (nucleotides) which are all essentially quite similar anyway. With proteins , 20 different amino acids, all with different sorts of properties, different sorts of charges, biochemical attributes, e.t.c, e.t.c. Therefore the world of proteins is very diverse and predicting the behavior of your molecule is much more difficult, still, protein work tends to be more rewarding , interesting and challenging, if you can get it to work at the end of the day.

Western blot is the important feature of this experiment. You have a slurry of proteins , i.e when you lyse a particular population of cells= millions of proteins. You want to see if a specific protein is present in this slurry and you are particularly interested in observing its molecular weight at the same time.

It works in two parts: The Gel part and the western part.

PART ONE: THE GEL PART.

First you need to run a gel often called polyacrylamide gel electrophoresis (PAGE). Acrylamide is a nasty nasty neurotoxin. Acrylamide +bis–acrylamide form the basis of the mesh–like structure. TEMED and ammonium persulfate actually catalyse the crosslinking. More specifically, TEMED and ammonium persulfate cause the production of free radicals resulting in crosslinking of your acrylamide fibres.

Amounts of TEMED + ammonium persulfate dictate rate of polymerization. Amounts of acrylamide + bis–acrylamide dictate pore sizes in mesh. Otherwise, an acrylamide gel runs on the same premise as agarose gels. Small things move easier through gel.

LAEMMLI SYSTEM: This set up is also called a discontinuous gel system which is used to make sure that you get nice, tight bands.

Stack: It has low percentage of acrylamide, pH 6.8 and chloride ions. The Chloride ions (Cl⁻) is the leading ion and moves faster than proteins.

Resolving: This is a set percentage of acrylamide, pH 8.8 , Chloride ions.

Running buffer: This contains glycine ions used as trailing ion.

Start gel, chlorine moves quickly . Glycine from buffer enters stacking gel and moves slowly. Proteins also move slowly, and are caught in an area in between the two different ions You can also envision that the chlorine ions move so fast , that there is an area of low conductance between the chlorine and the proteins.

This is why the proteins move slowly through the stack. When proteins reach boundary between stack and resolve , they see the resolving gel which has smaller pore sizes, making it tough for the protein to enter (this is also made worse by the fact that the chlorine ions are so far away that proteins aren't very attracted towards the positive electrode). But , when the slow glycine ion reaches the interface, it will become a dum, dum , dum fast glycine ion because of the different pH.

A result of all these is that all your protein sample will tighten into a sharp band at the boundary, and then get separated according to size . Proteins will be able to move into resolving because the glycine ion is pulling it towards the positive electrode.

Other things to keep in mind is that proteins are quite complicated in many sizes, many shapes , many variable charge properties. This will affect mobility. Consequently you need an equalizer. This is your sample buffer (the blue stuff).

- It has dye so that you can see the stuff.
- It has glycerol which makes sample heavy so that samples will flow into wells.
- It has SDS which is very important as it coats proteins with negative charge , (now all proteins have same charge density) , and denatures proteins to uniform shape (rod-like shape). Now all proteins have equivalent shape as well.
- It has beta- mercaptoethanol or DTT (dithiothreitol). These are reducing agents and will reduce and break disulfide bonds.

Anyway you run the gel, include protein standards which are set of predetermined molecular weight proteins. Show a picture of a gel, what a gel will look like after it is stained and dried.

Staining: Coomassie blue (0.3ug to 1ug per band) can probably see a band as faint as 100ng.

Silver staining: It is about 2-5ng and is much more sensitive very dirty procedure. If you want figure quality data, the equipment must be clean.

QUANTITATION OF PROTEINS:

- Run a gel and compare its band with preweighed amounts of standard protein like albumin . It works really well and is kind of labour intensive.
- Use colorimetric tests, BCA test works best and easy to do but most are very susceptible to chemicals that are commonly used in the buffers you store your proteins in (i.e. presence of detergents, tris, etc., etc.
- Conduct absorbance at 280nm, very rough and is approximately 1.0 O.D= 1mg/ml of protein. Very rough. And finally ;
- The bubble test.

PART TWO: THE WESTERN PART.

Now you do the western blot: This is the probing for specific protein using an antibody (ab). The main idea is that you cannot use a gel . It is like jello; fragile , sticky and proteins can diffuse readily through out the gel. So use something more convenient. Use a membrane and this is why it is called blot.

O/H PICTURE OF TRANSFER: So you have your gel and you have your membrane (usually nitrocellulose– don't prewet with methanol or PVDF). This membrane loves , loves, loves to bind proteins. **Proteins will transfer directly onto membrane such that all spatial positions of proteins are preserved.**

Now you have everything on membrane, easy to handle. Now these are the basic steps of the blot procedure: Use animated O/Hs.

STEP 1: Incubate membrane with block buffer , approximately 2 hours solution that covers the rest of the membrane with generic protein usually milk powder or BSA.

WASH: Wash membrane a bit approximately for 5 minutes. `

STEP 2: Incubate membrane with primary ab approximately 2 hours i.e. Ab that is specific for your protein of interest.

WASH: Wash several times ; approximately from 1 to 2 hours.

STEP 3: Incubate with secondary Ab for approximately 1 hour; Antibody that is specific for the Fc portion constant region of the primary Ab. It also has an enzyme attached to it that can convert some substrate into an observable product as a means of detection . i.e. it glows. It changes colour etc., etc.

WASH: Wash several times for approximately 1 to 2 hours.

STEP 4: Add substrate . Show western without the block step.

QUICK METHOD: This is done using Millipore methods. It uses PVDC membrane (immobulon P). Membrane is highly hydrophobic . ("Hates water") . Therefore , in theory, solutions can't touch membrane except in the areas that have protein attached. Proteins are not themselves extremely hydrophobic. Consequently, there is no need for the block step if membrane is dry (i.e. dry on bench for a minimum of one hour. Also actual surface area that solutions are working with is way less. Only places with protein count compared to previously where the entire membrane is protein covered.

Therefore, all steps are much shorter, primary Ab is 40 minutes to 1 hour, secondary Ab is 20 minutes to 30 minutes and washes are in seconds. The quick method works. I recommend trying it first in comparison to the normal technique and if you are convinced that the results are the same , then it is obviously a much more efficient protocol.

Some tricks associated with the technique include a quick 20 minutes block step , should your antibody be prone towards non-specific binding. Also, apparently you can speed up the drying process by immersing the membrane in 100% methanol for 10 minutes and then drying it for 10 minutes.

CHAPTER 22

RNA ANALYSIS TECHNIQUES.

RNA work is not fun. Although as a molecule, it is just as robust and sturdy as DNA (a little more even , since it can form some pretty fancy tertiary structures), it suffers because the nucleases (RNases) that can destroy it are very stable molecules. RNases are also everywhere . They don't require divalent cations as cofactors. Unharmed by autoclaving procedures or exposure to temperatures as high as 150 °C. Because of this , you have to take the following precautions: You must use aseptic technique , you quite seriously want to be as anal as possible. Segregate all your equipment, your space, even yourself , if you have to.

Different people exhibit different levels of care, but a general rule is to be as careful as possible. Remember, RNases are everywhere. So use disposal sterile plastic wear (polypropylene), and if you have to use glassware, treat it accordingly i.e. Glassware cleaned with detergent thoroughly , rinsed and over-baked at >250 °C for >4 hours. Show OH/DEC treated Barbie doll.

DEPC is a very useful reagent for making buffers and equipment RNase free. Essentially, you mix some DEPC with your buffer, let it sit, and get rid of the DEPC before using your liquid . For equipment you immerse it in a similar solution and let the DEPC evaporate away. What does it do? DEPC results in a covalent modification of nucleases rendering them inactive . Histidine modifier- imine > carbonate > NH > N–CO$_3$.

Things that are to be considered when using DEPC is that it comes as a liquid and you can buy at different percentages. It stinks. It is an organic solvent. Handy because you can tell its presence by its smell. Work in fume hood when treating water , buffers or glassware. Usually let your buffer +DEPC or equipment/glassware incubate for more than 12 hours. Then leave it in fume hood O/N with lid slightly ajar, so that fumes escape or autoclave the bugger if possible.

DEPC in contact with water creates CO_2 and ethanol. If not careful and kept in sealed container can lead to excessive pressure buildup. This is why you hear horror stories about it being explosive (not fireball explosive . It is more like glass shattering explosive).

If you have not heared of any incidents, just be careful when using it. You cannot treat Tris solutions with DEPC. Tris is full of amines that DEPC will spend more of its time modifying the wrong thing. When treating glassware or apparatus , immerse in fresh DEPC solution O/N for approximately 12 hours. You will know that it is there, if it stinks.

Then autoclave for 15minutes to remove residual DEPC. Alternatively, gel box cleaning can be cleaned with 0.5% SDS, rinsed with water, dried with ethanol, filled with 3% hydrogen peroxide for

10 minutes. Rinse thoroughly with RNase free distilled water (dH20).

Chloroform also denatures Rnases, rinse with chloroform. It can be quite harmful to plastic ware. GibcoBRL takes Rnase away . The reagent is $100 per 250ml. It can wipe apparatus clean. It is not harmful to plastics.

FINAL POINTERS: Always wear gloves. It is a good idea to change them periodically , may be go through 2 to 3 sets today. Once lysis has proceeded and RNA is exposed, keep things cold. RNases are not happy at 0.4 °C.

ISOLATION OF RNA: **General procedure**: We are extracting RNA from plant material using a product called Trizol (available from Gibco BRL) . Although this reagent is a proprietory product, it is most likely based on the familiar Phenol + Guanidium Thiocyanate procedure. Guanidium thiocyanate is often classed as a denaturant although it is also considered a chaotrophic salt (sucks water).

This method works in an analogous fashion to phenol chloroform extraction except that you also want to get your DNA to go into the phenol layer, therefore, you are left with just RNA in the aqueous phase. This Trizol business works because RNA is still water soluble in a high molar guanidium thiocyanate solution whereas proteins and DNA is not. Consequently, the insoluble components will tend to go to the organic phase.

ALTERNATIVE PROCEDURE: Another way of getting RNA is to utilize the fact that your RNA resides in the cytoplasm whilst the DNA resides in the nucleus , at least for eukaryotes. Consequently, an option is to first treat the cell with a "gentle" detergent to lyse the cell membrane but leave the nuclear membrane intact. Examples of common detergents used for this purpose are Triton X–100 and NP– 40. These two are almost identical).

NOTE: The biochemistry and behavior of detergents is very complicated. Consequently , when dealing with a detergent, it is a good idea to follow the procedures given rather than playing around too much. Detergents have many attributes that affect their effectiveness.

Dependent on their existence as free, form particles and micelle complexes (like a vesicle) . Detergent forming micelles don't work as well and micelle formation is very sensitive to both temperature and concentration effects which vary enormously from detergent to detergent.

NOTES REGARDING NORTHERNS: If you plan on transferring RNA over to a membrane , you have to go to extra efforts to get your RNA linear. RNA loves to form tertiary structures which will make sticking to the membrane difficult. So, if you use RNA , it is a good idea to include a denaturant step i.e . + formaldehyde, or use DMSO + glyoxal. (We have omitted this step , which may explain why it never works very well).

For RNA work, we used capillary action to transfer RNA that was treated with formaldehyde and run on a MOPS system gel. You cannot use Tris because of DEPC problem.

MORE STUFF ON RNA: RNA techniques are interested in RNA processing, mRNA amounts , what is in in vivo translation rate, and making a cDNA Library.

There is need to concern yourself with the quality of RNA preparation. Total RNA is approximately 80% rRNA, 15% tRNA and <5% mRNA. Most people are concerned about mRNA although some study rRNA as an evolutionary marker since the rRNA molecules are very conserved from species to species. They use it to guage evolutionary trees.

Generally, the cleaner the mRNA preparation, the easier your life is going to , be but getting purified mRNA sample is difficult in itself. Almost all mRNA have poly A tails which is a very useful characteristics that many mRNA techniques take advantage of. The problem is that Rnase degradation is an even bigger concern since, these nucleases tend to chew the ends of RNA first.

EXAMPLES OF USING THE POLY A TAIL: You can use oligo dT affinity chromatography to purify your mRNA. You can use the poly A region as a primer binding site for reverse transcriptase or PCR experiments.

The golden rule for mRNA work is as follows: Most techniques will still work with a total RNA preparation , but you will get much better results, if you go to the effort of purifying your mRNA first i.e. making cDNA library strongly suggests you use a purified mRNA prep. See HO of techniques, SI analysis, RNase protection and RT–PCR for mRNA quantitation.

SOME SPECIFICS ABOUT REVERSE TRANSCRIPTASE: Note that the procedure itself is very forward. Difficulties arise primarily because of the RNA prep. A number of variants of enzymes are available. AMV (Avian myeloblastosis virus), MMLV (Moloney murine leukemia virus), HIV (Human immunodeficiency virus), and other various proprietory enzymes. Some newer ones are also heat stable which means you can incubate at higher temperatures, therefore more stringent and more specific conditions. Unit designation is usually quite complicated and relates to the amount of incorporated labeled nucleoside in the reaction under specific temperatures and specific reaction times. Basically just follow instructions.

There are three basic functions; (1.) RNA –dependent DNA polymerase. (2.) Hybrid – dependent ribonuclease (RNaseH) and

(3.) DNA –dependent DNA polymerase. Oligo dT primer is usually a minimum of 12 nucleotides in length . It is commonly in the 12–20 nucleotide range. In prokaryotes, the common idea is to use random hexamer primer population.

TWO STEP AND ONE STEP RT–PCR: Two step is generally more reliable as you have the option of tweaking your PCR parameters (will talk on this later) i.e. you can make your PCR go at its optimal best , because the reaction can be fine tuned independently.

Sometimes you have to do two step because the RT you use prefers Mn^{2+} as a cofactor. There is more likelihood for cross contamination , since there are more steps. This can be a problem if your mRNA is in very low amounts and the sensitivity of your assay is high.

One step is quicker, less work but also less reliable because your PCR reaction conditions are restricted in the same conditions used in your reverse transcriptase assay, i.e. cofactor amounts stay the same. In the effect of your PCR primers in RT assay, etc. , etc., etc., one step reaction works because the two enzymes RT and the heat stable DNA pol work at different temperatures.

CHAPTER 23

PCR AND REVERSE HYBRIDIZATION ASSAYS

POLYMERASE CHAIN REACTION (PCR)

The polymerase chain reaction (PCR) is used for the direct detection of nucleic acids. It is particularly useful for the detection of viruses that are difficult to identify by conventional methods. The major requirement is that at least part of the sequence of the nucleic acid of the agent to be detected is known.

In essence , the technique is an in vitro technique for nucleic acid amplification specifically applied to a particular segment of DNA.

Two synthetic oligonucleotide primers are required that will hybridize to the ends of opposite strands of the target sequence. The process is cyclical involving three stages:

1. Denaturation of target DNA by heating at 90– 95 °C.
2. Cooling to 37 – 50 °C at which point the primers anneal to the target DNA strands and
3. Copying of the DNA and extension of the primers by means of a polymerase (Taq polymerase from *Thermus aquaticus* – A thermophilic bacterium) .

From 25 to 35 cycles can lead to a million–fold increase of the target DNA.

Today's basic premise is to look for a specific Alu insertion within the genomic sequence of TPA– 25 (Tissue plasminogen activator) . This insertion is very common in approximately 500 to 2000 per human genome. Some insertions are very recent which means ,in a given population , wide variance exist i.e. some will have it and some will not. We are going to be looking at one of these. Our insertion is in an intron (chromosome 8) and it is approximately 300bp in size. So , our task is to look up for a specific 300bp strand in a sequence of approximately 3000000000bp (no mean feat). We will be using PCR to do this.

PCR is an excellent example of illustrating how it is often the most simple and elegant ideas that really propel science or why the overwhelming majority of life science researchers believe that O.J Simpson is guilty. Garry Mullis, high on LSD looking at a fire and dreaming about snakes , came up with the idea of PCR , won a nobel prize in 1994 OR 1993 and shared it with Michael Smith at University of British Columbia.

The basic premise is that, working with DNA , if you want to manipulate it, or even see it, you need lots of it. PCR is a simple idea , but quite difficult to show pictorially.

KEY POINTS ABOUT PCR: Amplifications by rounds of replication in a test tube requires open or single -stranded DNA. Replication also needs a primase enzyme to make primer for the polymerase. But , you can make your own primer. Buy oligonucleotides. Now, we need nucleotides and can buy these , and we need a DNA polymerase.

With these constituents, we are all set. However, we do have one problem: this high temperature will basically knacker out any protein structure including our polymerase.

To get around this, let us use a polymerase from a bug that grows in high temperatures (thermophilic) i.e, this will be a heat stable polymerase.

The cycle is as follows:

- 95 °C denaturing temperature 1 min.
- X °C annealing temperature 1 min.
- 72 °C elongation temperature, 1- 3 min.

A picture of PCR reaction shows essentially doubling of the amount of DNA with every cycle. Therefore, after 30 cycles, you have approximately 1000000000 molecules of amplified product . That is a lot and therein lies the power of PCR.

You can get lots of products easily from very little material . Theoretically , one strand is all you need. But the true power of PCR is all the clever people who have tweaked with the procedure to do some very cool things. i.e.

- Site –directed mutagenesis.
- Production of restriction endonuclease site for convenient cloning.
- Production of single and double strand product for sequencing protocols.
- Quantitation of rare DNA.
- Amplification of partial cDNA sequences.
- Quantitation of mRNA expression.
- Differential display of mRNA by PCR.
- Anchored PCR to amplify regions upstream or downstream of known sequences.
- Random amplified polymorphic DNA (RAPD).
- Cloning dinosaurs from dinosaur blood found in fossilized mosquitoes trapped in amber.

WHAT YOU NEED FOR PCR:

1. General PCR buffer : salt + buffer + detergent. 50 mM KCl, 10mM Tris- Cl at pH 9.0, 0.01% Triton X-100 . Can make as a 10x buffer. This is simply a recipe that makes the polymerase.

2. You need DNA to amplify . This is the DNA you will start with. Approximately 1.0ug mammalian genomic DNA/ 10ul or 0.1 ng plasmid DNA/10ul use 10ul per reaction. The purer the better but PCR is quite forgiving.

3. You need primers each at 0.50 uM and 20- 30 nucleotides (nt) in length. The distance between two primers should be < 10 kb, but anything >3 kb will affect efficiency. Make it to complement the template exactly. If there are differences i.e. trying to introduce a restriction site, put the differences on the 5" end. GC content is similar to template sequence. Avoid polypurines, or polypyrimidines. Go over degenerate primers. This is quite commonly used but can lead to a lot of artifacts which of course, you have never heared about (not published e.t.c, e.t.c.) Proceed with caution, because you may waste a lot of time characterizing a PCR fragment that means nothing in reality!. Be warned against primer dimers.

4. You need dNTPS approximately 0.2 mM mix. Don't mess around , it is enough to make 12.5 ug of product!

5. You need Enzyme which usually comes in a stock of about 5 units /ul. You need approximately 2.5 units per reaction.

NOTE: Taq does this weird thing where an ATP is added to the 3" end when done.

THINGS TO TROUBLESHOOT WITH:

Generally, PCR is a standardized procedure but due to differences in primer and template, evitably, you may come across a PCR that won't go. So, when you are doing a particular PCR for the first time, it is worth figuring out the best conditions.

WHAT ARE THE CONDITIONS TO PLAY AROUND WITH?

1. **$MgCl_2$**: This is very important and it is probably the most important condition. The reason why is a little vague but Mg ions seems to do a number of things.
   - It binds DNA; may affect primer/ template interactions.
   - It binds taq polymerases ; required as a cofactor.
   - It influences taq polymerases ability to interact with primer/ template sequences.

The more the Mg, the less stringency in binding. It is a bit wavy as to exactly what is going on. Generally there appears to be4 no ground rules about what works best. So, if you are doing a PCR reaction for the first time, this is where you want to find out optimal Mg concentration. Do a titration , generally people will check between 1 and 6mM final concentration.

2. **+/- DMSO**: This has denaturant ability. It is good at keeping GC rich template / primer strands from forming secondary

structures. It doesn't seem to generally affect the reaction. Most people will include it regardless of that. Use at 5%.

3. **+/ – GLYCEROL**: This increases the apparent concentration of primer/template mix. Therefore may help in getting good primer/ template interactions at high temperatures . Use at 5%.

4. **ANNEALING TEMPERATURE AND ELONGATION TIMES.**
   There are a couple of guidelines: GC < 50%, use 55 ° C, if >50% use 60 ° C. OR use the ((A + T) x 2) + ((G + C) x 4) to figure out the Tm. Extension is at 72 °C . For < 500 nt, use 1 min. If greater than 500 nt, you may want to play around with time as high as 3min.

OPTIMISING THE FIRST ROUND OF AMPLICATION : It is very important for you to optimize the first round of amplication. The general idea is that whilst you are setting up your reaction, the taq pol , may start replication when the primer is binding at an inappropriate site. This will lead to "pseudo" bands and a DILUTING OUT EFFECT OF YOUR REAL TEMPLATE. So, there are a couple of tricks to optimize your first cycle.

- "hot start" + polymerase after first denaturation and annealing step , go to RT.
- Ice cooling after annealing temperature, go to ice temperature.
- Taq start antibody . Include addition of taq antibody which will get denatured upon 95 ºC step.

Finally check that the thermal cycler is working. Temperature needs to be spot on for this whole thing to work properly.

CHAPTER 24

LYSIS PROCEDURES.

OTHER LYSIS PROCEDURES:

1. (6hr): Use of CTAB detergent treatment: (cetyltrimethylammonium bromide). Compound forms and insoluble complex with nucleic acids (depending on NaCl concentration) . Most protocols optimized for bacteria or plant material.

2. (1 hr) : DTAB lysis quick protocol (good for 1hr preparation yielding approximately 10 ug DNA) Dodecyltrimethylammonium bromide. *Biotechniques 11, p298 (1991).*

3. (30 min) : DNAzol (Life technologies – Gibco BRL) . Special solution that contains guanidine isothiocyanate + special "proprietory" reagents. Instant lysis and purification step. Sample ready for precipitation. *Focus 18.1, p19 (1995).*

4. (<1 hr) : Modification of Glass beads and DNA extraction kits for blood samples . Good for small samples . Approximately 0.1 ug to 10ug DNA (i.e. dissolved quickly). Approximately works with both plant and animal tissue. *Biotechniques 15, p438. (1993).*

5. (15 min): Use of a microwave step, to aid in lysis of cell and extraction of DNA. Very small samples . Approximately 0.1ug to 10 ug DNA (i.e. can dissolve quickly) . Apparently works

with both plant and animal tissue. *Biotechniques 15, p438 (1993).*

6. (< hr) : Exploitation of high sedimentation coefficient of DNA , to separate away from RNA and protein. 15 min lysis steps. Tissue culture cells / blood preps . *Biotechniques 18, p413 (1995)*

7. (< 2 hr) : Combination of liquid nitrogen / proteinase K method. Physical breakage of frozen tissue. Lowers incubation time of proteinase K to 15 min. *Biotechniques 14, p163 (1993).*

NICK TRANSLATION ACTIVITY OF *ESCHERICHIA COLI* DNA POLYMERASE 1.

1. Duplex DNA.
2. Duplex DNA containing a nick with a 3" OH created by DNase 1.
3. One to several nucleotides removed from the 5' side of the nick by 5" – 3" exonuclease activity of *E. coli* DNA polymerase 1.
4. Excised nucleotides replaced by incorporation of labeled nucleotides by *E. coli* DNA polymerase 1.
5. Repetition of steps 3 and 4 result in translocation of the nick and uniform labeling of the synthesized DNA strand.

CHAPTER 25

USE OF PIPETTES AND THEIR TYPES

CHAPTER 26

BACTEC ™ MGIT ™ 960 SIRE. TEST FOR THE ANTIMYCOBACTERIAL SUSCEPTIBILITY TESTING OF *MYCOBACTERIUM TUBERCULOSIS* FOR FIRST AND SECOND LINE DRUGS.

CHAPTER 27

GT–BLOT 20 AUTOMATED EQUIPMENT FOR REVERSE HYBRIDIZATION ASSAYS.

CHAPTER 28

SPOLIGOTYPING: A PCR-BASED METHOD TO SIMULTANEOUSLY DETECT AND TYPE *MYCOBACTERIUM TUBERCULOSIS* COMPLEX BACTERIA FOR EPIDEMIOLOGICAL STUDIES: SPOLIGOTYPING AND SPOLIGOTYPING TECHNIQUES.

CHAPTER 29

MIRU–VNTR : MYCOBACTERIAL INTERSPERSED REPETITIVE UNITS – VARIABLE NUMBER TANDEM REPEATS; A STRAIN IDENTIFICATION AND DIFFERENTIATION PROTOCOL FOR THE EPIDEMIOLOGICAL STUDY OF *MYCOBACTERIUM TUBERCULOSIS.*

CHAPTER 30

MYCOBACTERIUM TUBERCULOSIS GENOTYPING TECHNIQUES

CHAPTER 31

BIOSAFETY AND LABORATORY SECURITY IN A MOLECULAR BIOLOGY CLINICAL REFERENCE LABORATORY.

PRINCIPLES OF BIOSAFETY AND LABORATORY SECURITY IN A TUBERCULOSIS AND HIV/AIDS REFERENCE LABORATORY.

This chapter examines the biological safety principles, characterization of four biosafety levels, areas of research needs , safety management programme profile, key tuberculosis and HIV/AIDS reference laboratory security protocols and the socio-economic implications of poor safety practices in a tuberculosis and HIV/AIDS reference laboratory.

Laboratory safety programmes are plans for preventing sickness and injury to personnel and damage or destruction of physical assets. The fundamental objectives of a meaningful laboratory safety programme include improvement of safety skills and attitude of all personnel , and development of a surveillance programme for prompt identification of hazards.

It also explains fully, the term "containment" which describes safe methods for managing infectious materials in the laboratory environment where they are being handled or maintained.

The purpose of containment is to reduce or eliminate exposure of TB/HIV/AIDS reference laboratory workers, other persons and the outside environment to potentially hazardous agents.

Primary containment, the protection of personnel and the immediate laboratory environment from exposure to infectious agents , is provided by both good microbiological techniques and the use of appropriate safety equipment to ensure biological safety and security in the tuberculosis and HIV/AIDS reference laboratories.

The use of vaccines may provide an increasing level of personnel protection. Secondary containment , the protection of the environment external to the Tuberculosis (TB) , Human immune deficiency virus (HIV), and Acquired immune deficiency syndrome (AIDS) reference laboratory from exposure to infectious materials, is provided by a combination of facility design and operational practice.

Therefore, the three elements of containment include laboratory practice and technique, safety equipment and facility design and also construction of secondary barriers.

The risk assessment of the work to be done with a specific agent will determine the appropriate combination of these elements.

BIOSAFETY LEVELS IN A TUBERCULOSIS AND HIV/AIDS REFERENCE LABORATORY.

Four biosafety levels (BSLs) are described in biomedical literature which consists of combinations of laboratory facilities. Each combination is specifically appropriate for the operations performed, and the laboratory function or activity.

The Laboratory Director is specifically and primarily responsible for assessing the risks and appropriately applying the recommended biosafety levels.

Biosafety level 1 practices.

Safety equipment and facility design as well as construction are appropriate for undergraduate and secondary educational training and teaching laboratories and for other laboratories in which work is done with defined and characterized strains of viable microorganisms not known to consistently cause diseases in healthy adult humans.

Bacillus subtilis, Naegleria gruberi , infectious canine hepatitis and exempt organisms under the NIH Recombinant DNA guidelines are representative of microorganisms meeting this criteria.

Biosafety level 1(P1) represents a basic level of containment that relies on standard microbiological practices with no special primary or secondary barriers recommended other than a sink for hand washing.

Biosafety level II practices.

Equipment, facility design and construction are applicable to clinical diagnostic , teaching and other laboratories in which work is done with the broad spectrum of indigenous moderate – risk agents that are present in the community and associated with human disease of varying severity.

With good microbiological techniques, these agents can be used safely in activities conducted on the open bench, provided the potential for producing splashes or aerosols is low. Hepatitis B Virus (HBV), HIV , the *Salmonellae* and *Toxoplasma* species are representative of microorganisms assigned to this containment level.

Biosafety level III (P2) is appropriate when work is done with any human– derived blood, body fluids, tissues or primary human cell lines where the presence of an infectious agent may be unknown e.g; Hepatitis and other highly infectious viruses, fungi and bacteria, TB/HIV/AIDS reference laboratories.

Tuberculosis and HIV/AIDS reference laboratory personnel working with human –derived materials should refer to the OSHA Blood borne pathogen standards for specific required precautions. 1

Primary hazards to personnel working with these agents include accidental percutaneous or mucus membrane exposures or ingestion of infectious materials. Extreme caution should be taken with contaminated needles or sharp instruments.

Even though organisms routinely manipulated at Biosafety level II are not known to be transmissible by aerosol route, procedures with aerosol or high splash potential that may increase the risk of such personnel exposure must be conducted in primary containment equipment or in devices such as biological safety or safety centrifuge cups.

Other primary barriers should be used as appropriate such as splash shields, face protection, gowns and gloves. Secondary barriers such as hand washing sinks and waste contamination facilities must be available to reduce potential environmental contamination.

Biosafety level III practices.

Equipment, facility design and construction are applicable to clinical , diagnostic , teaching, research or production facilities in which work is done with indigenous or exotic agents with a potential for respiratory transmission and which may cause serious and potentially lethal infection.

Mycobacterium tuberculosis ,that cause tuberculosis (TB), St. Louis encephalitis virus and *Coxiella burnetti* are representative of the microorganisms assigned to this level.

Primary hazards to personnel working with these agents relate to autoinoculation , ingestion and exposure to infectious aerosols in a TB laboratory.

Biosafety level III practices.

At Biosafety level III (P3) , more emphasis is placed on primary and secondary barriers to protect personnel in contagious areas, the community and the environment from potentially infectious aerosols. For example; all laboratory manipulations should be performed in a biological safety cabinet (BSC) or other enclosed equipment such as a gas–tight aerosol generation chamber.

Secondary barriers for this level include controlled access to the laboratory and ventilation requirements that minimize the release of infectious aerosols from the TB/HIV/AIDS reference laboratory.

Biosafety level IV practices.

Safety equipment, facility design and construction are applicable for work with more dangerous , exotic , foreign and strange agents that pose a high individual risk life – threatening diseases, which may be transmitted via the aerosol route ie. *Ebola virus,* SARS , *influenza, Marburg* and *Monkey pox virus* as well as multidrug and extensive drug resistant tuberculosis , and for which there is no available vaccine or therapy.

Agents with a close or identical antigenic relationship to biosafety level IV practice agents should also be handled at this level. When sufficient data are obtained, work with these agents may continue at this level or at a lower level.

Viruses such as *Marburg* or *Congo – Crimean haemorrhagic* fever or SARS are manipulated at biosafety level IV practice. The primary hazards to personnel working with Biological safety level IV agents are respiratory exposure to infectious aerosols , mucus membrane or broken skin exposure to infectious diagnostic materials, isolates and naturally or experimentally infected animals pose a high risk of exposure and infection to laboratory personnel, the community and the environment.

The laboratory worker's complete isolation from aerosolized infectious materials is accomplished primarily by working in a class III BSC or a full- body , air -supplied positive- pressure personnel suit.

The BSL 4 facility itself is a generally a separate building or completely isolated zone with complex, specialized ventilation requirement and waste management systems to prevent release of viable agents to the environment.

The Laboratory Director is specifically and primarily responsible for the operation of the laboratory. His/her knowledge and judgments are critical in assessing risks and appropriately represents those conditions under which the agents can ordinarily be safely handled.

Special characteristics of the agents used, the training and experience of personnel and the nature or function of the laboratory may further influence the director in applying these recommendations.

TB/HIV/AIDS reference laboratories and other clinical laboratories , especially those in health care facilities, receive clinical specimens with requests for a variety of diagnostic and clinical support services.

Typically, the infectious nature of clinical material is unknown and specimens are often submitted with a broad request for microbiological examination for multiple agents (e.g. sputa submitted for " routine" , acid – fast and fungal cultures) . It is the responsibility of the laboratory director to establish standard procedures in the laboratory, which realistically address the issue of the infective hazard for the director to establish standard procedures in the laboratory , which will realistically address the clinical specimens.

Except in ordinary circumstances, (e.g. suspected haemorrhagic fever) the initial processing of clinical specimens and serological identification of isolate can be done safely at BSL 2, the recommended level for work with blood borne pathogens such as HBV and HIV.

The containment element described in BSL IV (P4) are consistent with OSHA standard " Occupational safety and health administration (OSHA)). This requires the use of specific precautions with all clinical specimens of blood or other potentially infectious material (Universal or standard precaution) .

Additionally, other recommendations specific for clinical laboratories may be obtained from the American National Committee for Clinical Laboratory Standards (NCCLS).

The segregation of clinical laboratory functions and limited or restricted access to such areas is the responsibility of the laboratory director. It is also the Director's responsibility to establish standard written procedures that address the potential hazards and the required precautions to be implemented.

Selection of an appropriate biosafety level for work with a particular agent or animal study depends upon a number of factors.

Some of the most important are the virulence , pathogenicity, biological stability, route of spread and communicability of the agent, the nature or function of the laboratory, the procedures and manipulations involving the agent , the endemicity of the agent and the availability of effective vaccines or therapeutic measures.

LABORATORY SECURITY IN A TB/HIV/AIDS REFERENCE LABORATORY.

This is a very important aspect of laboratory management. The following guidelines were developed by the Centers for Disease Control and Prevention Atlanta, USA (CDC) 1 and adapted Vanderbilt Environmental Health and Safety (VEHS) 2 to address laboratory security for laboratories using biological agents or toxins capable of causing serious or fatal illness to humans or animals.

The guidelines are reflected below as follows:

Recognize that laboratory security is related but different to laboratory safety. This is done by involving both safety and security experts in evaluation and development of recommendation for a given facility or laboratory.

There should be review of safety and security procedures regularly. Management should review policies to ensure that they have good policies and procedures for both laboratory staff and visitors.

Laboratory supervisors should ensure that both laboratory workers and visitors understand the requirements and are trained and equipped to follow established procedures. Review and safety and security policies and procedures whenever an incident occurs or a new threat is identified should be regularly implemented.

Control access to areas where biologic agent or toxins are used and stored.

Laboratories and animal care areas should be separate from the public areas of the buildings in which they are located. Laboratory and animal care areas should be locked at all times. Card keys or similar keys should be used as permit to laboratory animal areas .

All entries (including entries by visitors , maintenance workers , repairmen and others needing one-time or occasional entry) should be recorded either by the card key device (preferable) or by signature in a log book.

Only workers required to perform a job should be allowed in laboratory areas and workers should be allowed only in areas and at hours required to perform their particular job. Access for students , visiting scientists etc. should be limited to hours when regular employees are present. Access for routine cleaning , maintenance and repairs should be limited to hours when regular employees are present.

Freezers, refrigerators, cabinets and other containers where stocks of biological agents, hazardous chemicals or radioactive materials are stored should be locked when they are not in direct view (eg. When located in unattended storage areas)

Know who is in the laboratory areas.

Facility administrators and laboratory directors should know all workers. Depending on the biological agents involved and the type of work being done , a background check and /or security clearance may be appropriate before new employees are assigned to the laboratory area.

All workers (including students , visiting scientists and other short-term workers) should wear visible identification badges and escorted or cleared for entry using the same procedures as for regular workers.

Know what materials are being brought into the laboratory area.

All packages should be screened (Visual or X-ray) before being brought into the laboratory area. Packages containing specimens, bacterial or virus isolates or toxins should be opened in a safety cabinet or other appropriate containment device.   Know what materials are being removed from the laboratory area. Biological materials and toxins for shipment to other laboratories should be packaged and labeled in conformance with all applicable local , federal and international shipping regulations.

Required permits should be in hand before materials are prepared for shipment. The recipient (preferably) or receiving facility should be known to sender and the sender should make an effort to ensure that materials are shipped to a facility equipped to handle those materials safely.

Hand carrying of microbiological materials and toxins to other facilities is rarely appropriate. If biological material or toxins are to be hand carried on common carrier, all applicable regulation must be followed.

Containment or possibly contaminated materials should be decontaminated before they leave the laboratory area. Chemicals and radioactive materials should be disposed of in accordance with local , state and federal regulations.

Have an emergency plan.

Control of access to laboratory areas can make an emergency response more difficult. This must be considered when emergency plans are developed. An evaluation of the laboratory area by appropriate facility personnel with outside experts if necessary to identify both safety and security concerns to be conducted before an emergency plan is developed.

Facility administrators, laboratory directors, principal investigators, laboratory workers , the facility safety office and facility security officials should be involved in emergency planning. Police, fire and emergency responders should be informed as to the types of biological materials in use in the laboratory areas and assisted in planning their responses to emergencies in the area.

Plan should include provision for immediate notification of (and response by) laboratory directors, laboratory workers, safety office personnel or other knowledgeable individual when an emergency occurs, so that they can deal with biosafety issues if they occur.

Laboratory emergency planning should be coordinated with facility-wide plans. Such facts like bomb threats, severe weather (hurricanes , floods) , earthquakes, power outages, and other natural or unnatural disasters should be considered when developing laboratory emergency plans.

Have a protocol for reporting incidents.

Laboratory directors, in cooperation with facility safety and security officials should have policies and procedures in place for reporting and investigation of incidents or possible incidents (e.g. undocumented visitors, missing chemicals, unusual or threatening phone calls etc.).

IMPLICATIONS OF POOR LABORATORY SAFETY AND SECURITY PRACTICES.

Laboratories are unique environments in relation to the safety of those who work within them. Clinical specimens received from patients pose a great deal of hazard to laboratory personnel because of the infectious agents that they may contain.

In addition to the threat of infection, laboratories also contain safety risks associated with any institutional environment namely ; those of fire, electrical and chemical hazards, hazardous environmental situations (slippery floors, faulty air- handling systems), radioactive materials, equipment malfunction and dangers imposed by natural disasters.

Laboratory safety programmes are plans for preventing sickness and injury to personnel and damage or destruction of physical assets. The fundamental objectives of a meaningful laboratory safety programme include improvement of safety skills and attitude of all personnel , and development of a surveillance programme for prompt identification of hazards. Others are formulation of plans for prompt correction of all hazards and co-ordination of laboratory safety effort with the overall laboratory safety and security programme ie. An employee health programme that must include pre- employment physical examination , with laboratory and radiological studies to establish fitness for laboratory employment.

This should be periodically repeated and employees should report all work-related illnesses and accidents. A general safety programme must include the orientation of new employees to department's attitude and policies for assuring safe laboratory conduct such as orderly housekeeping standards; storage and arrangement of supplies, eating , drinking, smoking and safe attire, safety officer and coordinated efforts for isolation of communicable diseases, control of nosocomial infections and plans for dealing with fire and other disasters.

A programme for handling chemicals which must include policies and labeling , transporting , storage, dispensing, and disposing of all chemicals. A programme for handling of all biological specimens which must include instructions for collecting , transferring, storage, processing and disposing of all specimens and specimen containers as well as instructions for hand washing and the care of work surfaces.

A fire prevention programme must include physical facilities and operational practices that satisfy fire code; handling and storing of combustibles, instructions for operating all heat- generating equipment (gas burners) ; and well -conceived and rehearsed plans in the event of fire e.g; sand buckets, fire extinguishers and fire blankets.

A first aid programme must include policies for dealing with all job-related injuries.

The health output of an individual leads to various social and economic consequences for that individual, the household, and the community as reflected in such measures as income, savings, physical and human capital, productivity , health consumption and social interaction.

These consequences are produced through a variety of response mechanism depending upon the severity of the disease and clinical manifestation. The difficulty of defining disease is implied in the very structure of the world "disease" . So many different kinds of disturbances can make a person feel not at ease and lead him to seek the aid of a physician that the word ought to encompass most of the difficulties inherent in the human conditions.

Generally, among the lay public, disease implies some serious organic and psychic malady such as cancer or insanity. Modern medicine in practice is broadening this concept to refer to any state, organic or psychic , real or imaginary , that disturbs a person's sense of well- being. In this sense , disease may threaten life or simply interfere with its enjoyment. It may prevent the sick person from functioning as a normal human being or simply from reaching his self-selected goals.

It is now realized that in dealing with the problems of the " disease" person, subjective and social factors may be as important as the objective organic lesions of behavioral disturbances recognized by the physician or psychiatrist.

Human history and our personal lives continue to be shaped by epidemics, ie. sudden outbreaks of infectious disease within a community . Many of the factors that determine the occurrence of disease in an individual also influence the spread of infectious diseases throughout a population.

Environmental conditions that bring organisms and hosts together encourage the spread of endemic disease. These environmental conditions include any physical, chemical, biological or social factors that are essential to the survival and transmission of the infectious agents. Only when these major elements– an infectious agent, a susceptible host and proper environmental conditions work in concert does infectious disease emerge and persist in a community.

The laboratory as an environment poses many hazards to unsuspecting and untrained people especially visitors. Epidemiology is the study of disease distribution. Social and cultural factors are important determinants of disease aetiology and distribution through their influence on the relationship between a human population and its natural environment, or through their direct influence on the relationship between a human population and its natural environment, or through their direct influence on the health of the population.

Social and cultural distinctions associated with differences in age, sex, occupation, class, ethnicity and community can have significant effects.

The incidence of numerous acute infections is highest in childhood, indicating that as people grow older, they develop immunities that decrease their vulnerability to these diseases. Death rates are clearly related to age. Obviously, these epidemiological patterns reflect biological variations in vulnerability to sickness and death associated with age differences .

It is therefore essential that good safety practices are employed to protect children and elderly relatives of laboratory personnel from laboratory- acquired infections.

The picture with respect to morbidity differences between the sexes is complete. But if male and female role distinctions can influence differences in response to illness, they can also influence differences in the development of illness as well; particularly if the culture emphasizes such role distinctions ; for example: house-keeping staff in the laboratories in Africa are usually females.

Studies of the effects of occupation on diseases have been an important part of epidemiology and has indicated that susceptibility to disease varied in accordance with means of livelihood.

In the laboratory, the unanticipated hazards of handling newly discovered, poorly understood, or previously un-encountered microbes exist alongside its myriads of problems. For example, the accidental infection of 31 researchers in Marburg , Germany in 1967 by a previously unknown virus in the tissues of African green monkeys is a remarkable event.

A substantial part of epidemiological research has been devoted to the influence of social stratification and ethnic differences on disease prevalence and aetiology.

This influence can be particularly significant in nutritional maladies and in certain infectious diseases whose spread is affected by the material conditions of life. For example, when people are crowded together indoors, respiratory infections (like common cold in schools , offices and meningococcal infections in military recruits) spread rapidly.

This is perhaps why respiratory infections are common in winter. The air in ill-ventilated rooms is also more humid , favouring survival of suspended microorganisms such as Streptococci and enveloped viruses.

A laboratory staff or visitor with laboratory associated infection may serve as an effective vehicle of transmission. Association of disease frequency with contrasting community setting , have formed another focus of epidemiological interest. Social correlates of rural – urban distinctions and their implications for health have been significant problems.

The role of patient, especially in many seriously disabling illnesses , may be extremely difficult. Even in so many cases, if a patient recovers or temporarily overcomes the effects of a deadly disease , he may not only regain his status in the community, but, even become an object of admiration. However, some diseases , in addition to actual or supposed impairment of the individual , carry with them an onerous burden of sigma; a social definition of disease that transforms the victim into a social outcast.

The mode of infection, does not alter the stigmatization, therefore, no one will admit having HIV infection , even if laboratory associated.

The prevention of illness and containment of disease are part of every medical system. But much more is involved than questions of sanitation, private and public cleanliness and robustness, for notions of contagion are bound up with religion and world view, and with perceptions of the powers and intentions of one's families , neighbors and friends; not to mention strangers.

Though intents may not be subject to control or change, behavior to some extent is , and therefore the medical system especially with regard to contagion and sanitation , is directly hooked into local systems of social organization and control.

Medical phenomena can be indicative of the performance of a social system. The health of its population is one significant test of effectiveness with which a society functions 3 .

One way to stay healthy and to insure the well-being of one's family , neighbors, friends and colleagues is to strictly observe all laboratory safety and security considerations. It is recommended that access to TB/HIV/AIDS laboratories should be limited to those who have been informed on how they can perform safely and securely.

Visitors , especially small children should be discouraged. Certain areas of TB/HIV/AIDS reference laboratories should be closed to visitors.

Finally, healthcare providers are expected to provide employees with all devices and mechanisms called engineering controls , necessary to protect them from hazards encountered during the course of work and in cases of accidents, immunization and first aid.

CONCLUDING REMARKS

The last several documented cases of small pox in the world, were acquired by employees working in a different area in the building where a small pox research laboratory was located. The principal Investigator was so dismayed that his possible deficient safety practices led to these small pox cases that he committed suicide.

It is the legal responsibility of the laboratory director and supervisor to ensure that an exposure control plan has been implemented and that the mandated safety guidelines are followed. Scientific exploration is not without hazards.

Each year , these explorers contract infectious diseases because of accidental exposure to pathogens in laboratories. Over 4,000 cases of laboratory associated infections have been recorded with more than 160 fatalities. The centers for disease control and prevention has defined which microbes currently represent the greatest hazard in the laboratory and has classified etiologic agents into four categories according to biosafety class levels.

Class 1 contains agents that pose little risk of serious disease. These include *Staphyllococcus epidermidis* and many other members of the normal flora.

Microbes in class 4 are the most dangerous and require the highest degree of containment. The plague bacillus is a class 4 -bacterium.

The greatest number of laboratory infections , has occurred among persons engaged in research activity.

Fewer than 23% of these infections have been reported as occurring in TB/HIV/AIDS diagnostic laboratories. This perhaps reflects the unanticipated hazard of handling newly undiscovered , poorly understood or previously un-encountered microbes.

For example, in 1967, 31 persons in Hamburg , Germany , as mentioned above, handling the tissues of African green monkeys were infected by a previously unknown virus in the tissue, and 6 of the victims died.

Most laboratory- associated infections are acquired by contact with infectious aerosols i.e. in the case of tuberculosis . Random air sampling demonstrates that common laboratory manipulations release microorganisms into the atmosphere e.g. blowing out pipette, removing stoppers or centrifugation.

Infected experimental animals or man may discharge contaminated respiratory droplets into the air, in addition, bites and scratches from these animals occasionally cause infections.

Mouth pipetting is one of the most hazardous of all laboratory manipulations, resulting in numerous cases of typhoid fever, tularemia, scarlet fever, hepatitis and influenza as well as other bacterial, parasitic and fungal opportunistic infections of HIV/AIDS.

Accidental inoculation with needles and syringes accounts for many laboratory associated infections such as HIV.

Spills of infectious materials are also implicated. In cases where a highly virulent pathogen is introduced into a susceptible population, an epidemic may occur. The affected population may be as small as a single family or as large as the global community.

CHAPTER 32

QUALITY CONTROL , QUALITY ASSURANCE AND EXTERNAL QUALITY ASSESSMENT IN A MOLECULAR BIOLOGY CLINICAL REFERENCE LABORATORY.

BASIC PRINCIPLES AND FOUNDATIONS OF QUALITY CONTROL, QUALITY ASSURANCE AND EXTERNAL QUALITY ASSESSMENT IN A TB/HIV/AIDS REFERENCE LABORATORY.

To ensure good laboratory practice in a Tuberculosis (TB), Human immune deficiency virus (HIV) and the Acquired immune deficiency syndrome (TB/HIV/AIDS) reference laboratory, there must be quality assurance (QA), quality control (QC), and External Quality Assessment program (EQAP). Quality assurance is the total process or activities that provides the evidence needed to give confidence that the quality activity is being performed properly.

Quality control is the measures that have to be included in each test run to verify that the test is working correctly. Quality control indicates whether a test run is valid. External quality assessment program is an external evaluation of laboratory performance based on participation in EQAP using proficiency panels.

The key issues for TB/HIV/AIDS laboratory quality assurance is that there must be detailed standard operating procedures (SOP) with total compliance checklists for monitoring all activities, organizational schemes for processing, documentation and assessment, monitoring of staff (competency evaluation) with blind proficiencies, neat and complete documentation of all results, no deviation from procedures, maintaining of confidentiality, endorsement of safety measures, and identification of errors and addressing them with a corrective action plan.

To have an effective quality plan, there must be focus on accuracy, detail, clarity and legibility. Check and recheck. Never assume. Verify and validate. There should be no short cuts. Follow standardized procedures and always focus on the objectives of the test in the TB/HIV/AIDS reference laboratory.

Quality Control (QC) in a TB/HIV/AIDS reference laboratory is a system of routine technical activities, to measure and control the quality of the test results as it is being developed. The QC system is designed to provide routine and consistent checks to ensure data integrity, correctness, and completeness, identify and address errors and omissions , document and archive inventory material and record all QC activities in the laboratory.

The following are the steps taken to ensure the QC in the tuberculosis reference laboratory; sterilization of TB/HIV/AIDS laboratory materials, TB/HIV/AIDS diagnostic laboratory tests are conducted to ensure that the cultures are not contaminated, regular fumigation of the environment of the laboratory is done, in order to destroy the pathogens in the air and environment of the laboratory, periodic test of the cultures to detect contamination by chemical or microorganism and steps to insure that the Lowenstein – Jensen slopes in the universal containers are sterile and does not have leakages that will encourage contamination.

Quality Assurance (QA) activities carried out in the tuberculosis reference laboratory include a planned system of review procedures conducted by personnel not directly involved in the inventory compilation/development process. Reviews, preferably by independent third parties, should be performed upon a finalized inventory following the implementation of QC procedures. Reviews verify that data quality objectives were met, ensure that the inventory represents the best possible estimates given on the current state of scientific knowledge and data available, and support the effectiveness of the QC programme.1

QUALITY CONTROL IN A TB/HIV/AIDS REFERENCE LABORATORY

Quality control in a TB/HIV/AIDS reference laboratory is an indispensable part of an effective TB/HIV/AIDS control programme. It encompasses the whole process of sample collection, processing , recording and reporting. The purpose of quality control programmes is the improvement of the efficiency and reliability of TB/HIV/AIDS reference laboratory services.

A quality control programme in a TB/HIV/AIDS reference laboratory is a process of effective and systematic internal monitoring which aims to detect the frequency of errors against established limits of acceptable test performance in the TB/HIV/AIDS reference laboratory.

Although it is not usually feasible to determine error frequency accurately, it is nevertheless a mechanism by which TB/HIV/AIDS reference laboratories can at least validate the competency of their TB/HIV/AIDS diagnostic services.

Quality control in the TB/HIV/AIDS reference laboratory also includes proficiency testing, which is also known as external quality assessment. This programme is designed to allow TB/HIV/AIDS reference laboratories to assess their capabilities by comparing their results with those obtained with the same specimens in other TB/HIV.AIDS laboratories of the network, e.g. regional , national or suparanational tuberculosis reference laboratories.

Quality control in a TB/HIV/AIDS reference laboratory also includes quality improvement. Quality improvement in a TB/HIV/AIDS reference laboratory is a process by which the components of domestic services are analyzed with the aim of looking for ways to permanently remove obstacles to success.

Data collection, data analysis , identification of problems and creative problem solving are the key components of this process. It involves continued monitoring and identification of defects , followed by remedial action to prevent recurrence of problems.

QUALITY CONTROL AND LABORATORY ARRANGEMENT AND ADMINISTRATION IN A TB/HIV/AIDS REFERENCE LABORATORY.

Quality control measures must be in place in TB/HIV/AIDS reference laboratory arrangement and administration. We must ensure that doors in the laboratory are always closed. Work areas , equipment and supplies should be arranged for logical and efficient work flow.

Work areas should be kept free of dust. Benches should be swabbed a least once a day with an appropriate disinfectant (e.g. 5% Phenol).

Every procedure performed in the TB/HIV/AIDS reference laboratory must be written out exactly as carried out and be kept in the TB/HIV/AIDS reference laboratory for easy reference. All records should be retained for two years. TB/HIV/AIDS reference laboratory procedures used routinely should be those that have been published in reputable medical books, manuals or journals.

QUALITY CONTROL AND LABORATORY EQUIPMENT IN THE TB/HIV/AIDS REFERENCE LABORATORY.

Equipment in the TB/HIV/AIDS reference laboratory should meet the manufacturer's claims and specifications. Written operating and cleaning instructions must be kept in a file for all equipment . TB/HIV/AIDS reference laboratory equipment must be monitored regularly to ensure the constant accuracy and precision necessary.

QUALITY CONTROL AND SPECIMENS /REQUEST FORMS IN A TB/HIV/AIDS REFERENCE LABORATORY.

TB/HIV/AIDS tests must be performed only upon written request from authorized persons and we do not allow oral requests without follow-up written instructions. We should insist on specimen request forms being kept separate from the specimens themselves.

Forms that has been contaminated by specimens should be sterilized by autoclaving. We should insist on adequately completed request forms and proper labeling of specimens to ensure positive identification of patient.

We should reject specimens that cannot be properly identified. We should evaluate the quality of sputum/blood samples and make a note if a specimen is of low quality or not suitable in the case of saliva for *Mycobacterium tuberculosis* identification.

The report should state, "Specimen resembled saliva- treat a negative result with caution" . To facilitate reporting , a rubber stamp of the comment can be made. We should destroy and discard leaking or broken specimen containers by autoclaving and request for repeat specimen.

We should also document the arrival time of specimens in the TB/HIV/AIDS reference laboratory and note any delays in delivery on the report form, particularly with negative or contaminated results.

QUALITY CONTROL AND REAGENTS/STAINS IN A TB/HIV/AIDS REFERENCE LABORATORY.

All containers of stains and reagents should show the date received and the date first opened. Any material found to be unsatisfactory should be recorded as such and removed from the TB/HIV/AIDS reference laboratory immediately. Stocks should be limited to six months' supply and regular stock rotation should take place to avoid unnecessary expiry.

QUALITY CONTROL AND BIOCHEMICAL TESTS IN A TB/HIV/AIDS REFERENCE LABORATORY.

The reagents in the TB/HIV/AIDS reference laboratory should be prepared as indicated and the expected biochemical test response checked by using appropriate positive and negative controls.

QUALITY CONTROL AND BIOSAFETY CABINETS IN A TB/HIV/AIDS REFERENCE LABORATORY.

In a TB/HIV/AIDS reference laboratory, the biological safety cabinet (BSC) is the primary containment device that protects the worker, product and /or environment from TB/HIV/AIDS and its performance needs to be verified at the time of installation and annually thereafter.

BSC Class 1 : In the Class 1 BSCs, unfiltered room air is drawn across the work surface.

Staff protection is provided by this inward flow as long as a minimum velocity of 75 linear feet per minute (22.8 meter per second) is maintained through the front opening.

Any airborne bacteria are entrained and conveyed into the HEPA filter. The class 1 BSC is hard-ducted to the building exhaust system and the building exhaust fan provides the negative pressure necessary to draw room air into the cabinet.

Modern cabinets have airflow indicators and warning devices. The filters must be changed when the airflow falls below the minimum velocity level.

BSC Class 2: In the Class 2 BSCs, the BSC provide staff, environmental and product protection. In BSC Class 2, airflow is drawn around the operator in the front grille of the cabinet, which provides staff protection.

In addition, the downward laminar flow of HEPA- filtered air provides product (sample) protection by minimizing the chance of cross-contamination along the work surface of the cabinet. Because cabinet air has passed through the exhaust HEPA filter, it is contaminant - free and may be circulated back into the laboratory BSC Class 1 or ducted out of the building class 2 biological safety cabinet (BSC).

Presently, **BSC Class 3 and BSC Class 4** are mostly used in TB/HIV/AIDS reference laboratories.

The purpose and acceptance level of the performance tests are to ensure the balance of inflow and exhaust air, the distribution of air onto the work surface and the integrity of the cabinet.

Other quality control tests are used to check the electrical and physical features of the BSC and these checks are done daily and they include that we should ensure that the rate of airflow across the front opening is 75 linear feet/minutes (22.86 meter/ second) for class 1 and 75 to 100 linear feet/minute (22.86 to 30.48 meter/second) for class 2 cabinets.

We should check also the magnetic guage in the exhaust duct for any pressure drop across the filters and replace the filters when the guage indicates that the airflows across the front opening has dropped below optimal levels.

The following tests should also be performed annually on class 1 and class 2 cabinets to ensure quality control in the TB/HIV/AIDS reference laboratory .

Downward velocity and volume test.

This test is performed to measure the velocity of air moving through the cabinet work space.

Inflow velocity test

This test is performed to determine the calculated or directly measured velocity through the work access opening, to verify the nominal set point average in flow velocity and to calculate the exhaust airflow volume rate.

An electric vane type anemometer should be used to measure airflow. The airflow into a class 1 cabinet should be measured in at least five places in the plane of the working surface and an average calculated.

At no place should there be a reading that is 20 linear feet/minute (0.1meter/ second) more or less than any of the others. If there is such a difference, there will be turbulence within the cabinet.

In class 2 cabinets, the airflow is greater at the bottom than at the top of the working space. The average inward flow is calculated by measuring the velocity of air leaving the exhaust and the area of the exhaust vent. From this, the volume per minute is calculated, which is also the amount entering the cabinet divided by the area of the working face , to give the average velocity.

The downward velocity of air should be measured at 18 points in the horizontal place, 10 cm above the top edge of the working face.

No reading should differ from the mean by more than 20%.

Airflow smoke pattern tests

This test is performed to determine if the airflow along the entire perimeter of the work access opening is inward, if airflow within the work area is down ward with no dead spots of refluxing, if ambient air passes onto or over the work surface, and if there is refluxing to the outside at the window wiper gasket and side seals.

The smoke test is an indicator of airflow direction not of velocity.

Commercial airflow testers are recommended. They are small glass tubes , seated at each end . Both ends are broken off with the gadget provided and a rubber bulb filter to one end.

Pressing the bulb to pass air through the tube causes it to emit white smoke.

HEPA filter leak test.

This test is performed to determine the integrity of supply and exhaust HEPA filters, filter housing, and after- mounting frames while the cabinet is operated at the normal set point velocities.

An aerosol is the form of generated particulates of dioctylph-thalate (DOP) or an accepted alternative is required for leak-testing HEPA filters and their seals. Although DOP has been identified as a potential carcinogen, competent service personnel are trained to use these chemical in a safe manner.

The aerosol is generated on the intake side of the filter, and particles passing through the filter or round the seal are measured with a Photometer on the discharge side. This test is suitable for ascertaining the integrity of all HEPA filters.

Cabinet leak test.

The pressure holding test is performed to determine if exterior surfaces of all plenums, welds , gaskets and plenum penetrations or seals are free of leaks.

It is performed just prior to initial installation when the BSC is in a free-standing position in the room in which it is used, after a cabinet has been relocated to a new location, and again after removal of access panels to plenums for repairs or a filter change. This test may also be performed on fully installed cabinets.

Electrical leakage and ground circuit resistance and polarity test.

These safety tests are performed to determine if a potential shock hazard exists by measuring the electrical leakage, polarity ground fault interrupter function and ground circuit resistance with the cabinet connection.

The polarity of electrical outlets are checked using a polarity tester. The ground fault circuit interrupter should trip when approximately 5 milliampere (ma) is applied.

Light intensity test

This test is performed to measure the light intensity on the work surface of the cabinet as an aid in minimizing cabinet operator's fatigue.

Vibration test

This test is performed to determine the amount of vibration in an operating cabinet as a guide to satisfactory mechanical performance, as an aid in minimizing cabinet operator's fatigue, and to prevent damage to delicate tissue culture specimens.

Noise level test.

This test is performed to measure the noise levels produced by the cabinet, as a guide to satisfactory mechanical performance and as an aid in minimizing cabinet operator's fatigue.

QUALITY CONTROL AND CENTRIFUGE IN THE TB/HIV/AIDS REFERENCE LABORATORY:

We should check the brushes and bearings every 6 months.

QUALITY CONTROL AND INCUBATOR 35 o C – 37 o C IN THE TB/HIV/AIDS REFERENCE LABORATORY.

We should record the temperature daily, preferably in the morning. Test the temperature at several sites within the incubator by placing a thermometer in a water reservoir (e.g. Erlenmeyer Flask). Control the light within the incubator by covering the glass front of the incubator door and by restricting the use of any lights inside the incubator.

QUALITY CONTROL AND INSPISSATOR IN THE TB/HIV/AIDS REFERENCE LABORATORY.

Check temperature daily, clean after each batch of culture medium prepared.

QUALITY CONTROL AND PH METER IN THE TB/HIV/AIDS REFERENCE LABORATORY.

Compensate for temperature with each run. Date buffer solutions and discard when unsatisfactory. Standardize with pH 4.0 and 7.0 buffers before each test or sense of tests.

QUALITY CONTROL AND WATER BATHS IN THE TB/HIV/AIDS REFERENCE LABORATORY.

We should check temperature before and during the use of water bath. We should clean it monthly.

QUALITY CONTROL AND REFRIGERATOR 2 ° C– 8 ° C IN THE TB/HIV/AIDS REFERENCE LABORATORY.

We should check the temperature daily . Clean it monthly. Defrost or check refrigerator and freezer compartment every 3 months.

QUALITY CONTROL AND FREEZERS IN THE TB/HIV/AIDS REFERENCE LABORATORY.

We should check it daily and clean it every 6 months.

QUALITY CONTROL AND GLASSWARE IN THE TB/HIV/AIDS REFERENCE LABORATORY.

We should discard chipped or etched glassware. We should ensure that glassware are free of detergents. We should not store sterile glassware for more than three weeks before it is used.

QUALITY CONTROL AND DIGESTION / DECONTAMINATION IN THE TB/HIV/AIDS REFERENCE LABORATORY.

We should process sputum specimens in batches according to centrifuge capacity. We should keep a monthly record of the percentage of clinical specimens contaminated ; acceptable range is 2-5% .

Contamination rates <2% indicates overly harsh decontamination, which means that too many tubercle bacilli are killed. If the laboratory is experiencing delays in delivery of specimens , the contamination rate may be greater than 5%. If the rate of >5% persists, ensure that specimens are completely digested , since partially digested specimens may not be completely decontaminated.

Thoroughly mix the contents of the centrifuge tubes to ensure that the inside surfaces have been well decontaminated.

QUALITY CONTROL AND CULTURE MEDIA IN THE TB/HIV/AIDS REFERENCE LABORATORY.

We should use fresh eggs (<seven days) for preparation of Lowenstein - Jensen media. We should also control coagulation time and temperature for egg-based medium. We should discard media that are discoloured or have bubbles following inspissation. We should check all batches of media for sterility by inoculation at 35 o - 37 o C for 24 hours. We should keep all media in the dark in the refrigerator and discard unused media after four weeks.

QUALITY CONTROL AND CULTURE PROCEDURES IN THE TB/HIV/AIDS REFERENCE LABORATORY. We should avoid cross- contamination of cultures using individual pipettes or loops and strict aseptic techniques. We should be suspicious of several successively positive specimens or cultures with few colonies that follow a heavily positive culture.

QUALITY CONTROL AND WATER IN THE TB/HIV/AIDS REFERENCE LABORATORY.

We should check both distilled water and tap water regularly for the presence of acid- fast contaminants. If water appears cloudy or dirty , centrifuge 200 – 250 ml in multiple tubes and make a smear of the combined sediment.

Alternatively , filter 1000 ml of water through a sterile 0.22 Um pore size membrane filter. Cut them into strips with a sterile scissors and place on Lowenstein – Jensen culture medium.

CONCLUDING REMARKS.

Quality control in a TB/HIV/AIDS reference laboratory is a system designed to continuously improve the reliability , efficiency and use of TB/HIV/AIDS reference laboratory services as a diagnostic and monitoring option.

The purpose of the quality control programme is to improve the efficiency and reliability of TB/HIV/AIDS reference laboratory services. This includes a process of effective and systematic internal monitoring of the performance of bench work in the TB/HIV/AIDS reference laboratory.

Quality control ensures that the information generated by the laboratory is accurate, reliable and reproducible. This is accomplished by assessing against acceptable established limits; the quality of specimens, the performance of decontamination and processing procedures, the quality of reagents, media and equipment, by reviewing results and by documenting the validity of laboratory methods.

Quality control should be performed on a regular basis in a TB/HIV/AIDS reference laboratory to ensure reliability and reproducibility of laboratory results.

For a quality control programme to be of value, it must be practical and workable. Quality control must be applied to laboratory arrangement, equipment, collection and transportation of specimens, handling of specimens, reagents and media, sample processing methods and reporting of results.

The keys to successful quality control are; adequately trained, interested and committed staff, common-sense use of practical procedures, a willingness to admit and rectify mistakes and effective communication.

In further discussion, on ensuring a reliable and quality laboratory service, a reliable and quality laboratory service is achieved and sustained not just by implementing quality control of laboratory tests. This is important but only part of what is needed.

Increasingly, the term total quality management (TQM) is being used to describe a more comprehensive and user-orientated approach to quality.

TQM addresses those areas of laboratory service that most influence how a laboratory service functions and uses its resources to provide a quality and relevant service.

TQM in laboratory practice includes correct use of the laboratory, providing a quality service to patients and those requiring tests, management of finances, equipment and supplies, staffing of laboratories, training and competence of staff, quality assurance to obtain correct test results, responsibility for TQM and continuing improvement in quality.

TQM incorporates both the technical aspects of quality assurance and those aspects of quality that are important to the users of a laboratory service, such as information provided, its correctness and presentation, time it takes to get a test result, and the professionalism and helpfulness of laboratory staff.

Such a comprehensive commitment to quality is essential to achieve the best possible service to patients, user confidence, effectiveness and efficiency, accountability and optimal use of resources.

Successful total quality management (TQM) of laboratory services requires close collaboration between laboratory staff, those who request laboratory tests, laboratory coordinator, hospital medical officers and the health management team.

CHAPTER 33

GOOD LABORATORY PRACTICES AND STANDARDS IN A MOLECULAR BIOLOGY CLINICAL REFERENCE LABORATORY.

PRINCIPLES AND FOUNDATIONS OF GOOD LABORATORY PRACTICE AND STANDARDS IN A TUBERCULOSIS AND HIV/AIDS REFERENCE LABORATORY.

This chapter examines the principles of the basic standard practices required in a Tuberculosis and HIV/AIDS Reference Laboratory. This explains the minimum standard of practice required by every Tuberculosis and HIV/AIDS Reference Laboratory in the day to day running of its affairs. Adhering to these principles and standards would ensure that elements of good laboratory practice and standards in a Tuberculosis and HIV/AIDS Reference Laboratory like confidentiality, counseling, quality control, quality assurance, quality assessment, standard operational procedure, supervision, safety precaution, water requirement , reagents and kits are practiced in the Tuberculosis and HIV/AIDS Reference Laboratory for the good clinical management of the patients.

Safety in the Tuberculosis and HIV/AIDS Reference Laboratory is a part of good laboratory practice (GLP). The formal concept of Good Laboratory Practice (GLP) evolved in the USA in the 1970s due to concerns about the validity of pre- clinical safety data submitted to the food and drug administration (FDA) in the context of new drug applications (NDA) .

The inspection of studies and test facilities had implications for instances of inadequate planning and incompetent execution of studies, insufficient documentation of methods and results and even fraud.

Good laboratory practice is defined as a quality system concerned with the organizational process and the conditions under which health and environmental safety studies are planned, performed, mentioned , recorded , archived and reported. The purpose of these principles of GLP is therefore to promote the development of quality test data, to provide a management tool to ensure a sound approach to management including conduct, reporting and archiving of laboratory studies. The principles may be considered as a set of criteria to be satisfied as a basis for ensuring the quality , reliability and integrity of the studies, the reporting of verifiable conclusions, and the traceability of data.

The elements of good laboratory practice and standards in a Tuberculosis and HIV/AIDS Reference Laboratory include confidentiality, counseling, quality control, quality assurance, quality assessment, standard operational procedure, supervision, safety precaution, water requirement, reagents and kits. Let us examine each of them further.

CONFIDENTIALITY IN A TUBERCULOSIS AND HIV/AIDS REFERENCE LABORATORY

Confidentiality are possible structures put in place to ensure confidential handling of test results in a clinical microbiology laboratory. Specific laboratory results and information on individuals tested should never be a topic of loose discussions. The privacy and rights of an individual can be severely compromised by information from overhead conversations. Test results should not be available for general viewing. Results must be kept in a secure location in a sealed envelope marked " Confidential". Then hand delivered to the submitting Doctor to maintain confidentiality.

Results should not be communicated via telephone or e-mail to avoid break in confidentiality.

COUNSELING IN THE TUBERCULOSIS AND HIV/AIDS REFERENCE LABORATORY.

This is the pre and post- test counseling of patient's management and the role of laboratory scientists in ensuring that patients are prepared for their test results before disclosure. This is very important in TB/HIV/AIDS infection management. A counseled patient is able to handle and live positively with a positive result.

No individual should be tested without having received pretest counseling. A pretest counseling session would explore client's knowledge of the disease and correct myths and misconceptions, inform the client about the "what" and "how" of testing , assess risk behavior, history of the disease, blood transfusion, drug injections , sex work, bisexuality, multiple sex partners, non -protective sex, skin cutting and piercing procedures, history of sexually transmitted infections, the meaning and implications of a positive or negative test result must be exposed with each patient and also ask client who he/she might want to inform, if the test is positive.

Results are best handled by sending them back to the Doctor or Counselor who requested for them rather than handing the result back to the patient.

This would ensure that the patient receives post -test counseling once the result is disclosed. Post- test counseling session would include all clients that come in to get their results personally whether negative or positive and inform the client of their result , if they are ready for it.

For a positive result, some sensitivity is required in giving the result. Allow client to vent emotions and help client come to terms with the result.

This process of coming to terms with the result may take several weeks or months. Re-inform clients about the availability of drugs , what it can do and the implications for commencing.

For a negative result, inform client about window period and the need to repeat the test in another 12 weeks. Do not rejoice over a negative result or weep over a positive result.

QUALITY CONTROL IN THE TUBERCULOSIS AND HIV/AIDS REFERENCE LABORATORY.

Quality control (QC) are all the issues related to ensuring sample integrity and maintaining controls and standards. Quality control refers to those measures that must be included during each assay to verify that the test is working properly.

The quality control programme in a tuberculosis and HIV/AIDS reference laboratory must be clearly defined and well- organized. The QC program must provide the system design and evaluation of proper patient identification and preparation which includes specimen collection, identification and preservation, specimen transportation , specimen processing and accurate result reporting.

This system must ensure optimum patient specimen and result integrity throughout the pre-analytical , analytical , and post-analytical processes.

Opportunities for system improvement are identified and based on such evaluations. Judgement of the acceptability of QC data must be made before patient results are reported. Oversight review must occur at least monthly by the Laboratory Director or designee. Beyond these specific requirements, a laboratory may (optionally) perform more frequent review at intervals that it determines appropriate for its setting and the assays involved.

Parts of quality control in the tuberculosis and HIV/AIDS reference laboratory include:

Sample integrity in the tuberculosis and HIV/AIDS reference laboratory.

There should be documented criteria for the rejection of unacceptable specimens and the special handling of sub–optimal specimens. There must be a mechanism to notify the requesting Physician and discuss whether testing is appropriate for that specimen. Age and sex specific reference intervals (normal values) must be verified or established by tuberculosis and HIV/AIDS Reference laboratory.

If a formal reference interval study is not possible or practical, then the laboratory should carefully evaluate the use of published data for its own reference ranges, and document that review. All reagents must be properly labeled as applicable and appropriate, with the elements of content and quality , concentration or titer, storage requirements, date prepared or reconstituted by the laboratory and expiration date.

There is no requirement to routinely label individual containers with " date opened" , however , a new expiration date must be recorded on the container , if opening the container changes the expiration date or storage requirement.

The laboratory must assign an expiration date to any reagent that do not have a manufacturer- provided expiration date based on known stability, frequency of use, storage conditions, and risk of contamination. Reagents must not be used beyond their stated or assigned expiration date.

Reagents must be stored as recommended by the manufacturers in order to prevent environmentally- induced alterations that could affect test performance.

If ambient temperature is indicated, there must be documented that the defined ambient temperature is maintained and corrective action is taken when tolerance limits are exceeded. New reagent lots should be checked against old reagent lots or with suitable reference material before or concurrently with being placed in service.

For qualitative tests, minimum cross-checking includes retesting at least one known positive and one known negative patient sample from the old reagent lot against the new reagent lot, ensuring that the same results are obtained with the new lot.

Commercial reagents and controls must be used according to the manufacturer's instructions. If alternative procedures are used, the method must be evaluated to justify the change. If there are multiple components of a reagent kit, the laboratory must use components of reagent kits only with other kits that are in the same lot number, unless otherwise specified by the manufacturer.

Controls and standards in the tuberculosis and HIV/AIDS reference laboratory.

Controls are samples that act as surrogates for patient specimens. They are periodically processed like a patient sample to monitor the ongoing performance of the entire analytic process.

The laboratory is expected to provide validation of all instrument - reagent systems for which daily controls are limited to including instruments and electronic controls. It is implicit in quality control that control specimens are tested in the same manner as patient specimens.

Moreover, QC specimens must be analyzed by personnel who routinely perform patient testing. To the extent possible, all steps of the testing process must be controlled. For qualitative and semi-quantitative antigen- antibody tests, that do not include a positive and negative external control, must be tested with each new kit lot number or different shipment of a given lot number in the laboratory.

Temperature of storage equipment in the tuberculosis and HIV/AIDS reference laboratory.

The temperature of water baths or heating blocks , refrigerators, freezers and other temperature- dependent equipment containing reagents and patient specimens must be monitored daily , as equipment failures could affect accuracy of patient test results.

Items such as water baths and heat blocks used for procedures need only be checked on days of patient testing. Automatic and adjustable pipetting devices must be checked at specified period intervals for accuracy and reproducibility and the results of such testing documented.

Accuracy of volumetric glassware in the tuberculosis and HIV/AIDS reference laboratory.

Volumetric glassware must be certified for accuracy or checked for accuracy before being placed in service. Disposable micropipettes must be examined visually for uniformity and length of column and a representative sample checked before the box is placed in service.

Data management in the tuberculosis and HIV/AIDS reference laboratory.

The laboratory must have a documented system in operation to detect and correct clerical and analytical errors that could affect patient management.

The laboratory must have a documented system in operation to verify highly unusual results for each test or instrument. There must be a protocol for review of highly unusual patient data by an experienced individual when such data are used for patient management decisions.

The system for detecting clerical errors, significant analytical errors and unusual laboratory results must provide for timely correction of errors.

Equipment in the tuberculosis and HIV/AIDS reference laboratory.

There must be evidence of documentation of corrective actions taken when instrument function , temperature etc. ,exceed defined tolerance limits. The laboratory must calculate precision statistics (such as standard deviation (S.D)) at specified intervals for numeric quality control data. The quality improvement programme must cover all aspects of the laboratory service.

Key indicators of quality must be monitored regularly and evaluated for opportunities to improve patient care. There must be evidence that the chosen indicators are being compared against a benchmark , where available and applicable. The benchmark may be a published practice guideline or the laboratories' own experience (Trend analysis).

QUALITY ASSURANCE IN A TUBERCULOSIS AND HIV/AIDS REFERENCE LABORATORY.

Quality assurance of specimen includes quality assurance of laboratory supply , quality assurance of analysis , measures of checking indeterminate errors and external quality assurance in the tuberculosis and HIV/AIDS reference laboratory.

QUALITY ASSESSMENT IN THE TUBERCULOSIS AND HIV/AIDS REFERENCE LABORATORY.

This is also known as proficiency testing. It is a means of determining the quality of results generated by the laboratory. It is usually an external evaluation of a laboratory's performance , involving the incorporation of proficiency panels as the means of evaluation.

Quality assessment measures the effectiveness of the quality control and quality assurance.

STANDARD OPERATING PROCEDURE   IN THE TUBERCULOSIS AND HIV/AIDS REFERENCE LABORATORY.

The tuberculosis and HIV/AIDS reference laboratory must have a completely documented procedure describing methods for patient identification, patient preparation , specimen collection and labeling, specimen preservation, conditions for transportation and storage before testing.

Such protocols must be consistent with good laboratory practice and openly displayed for easy sighting and referencing.

SUPERVISION   IN THE TUBERCULOSIS AND HIV/AIDS REFERENCE LABORATORY.

The supervisory or inspection team should review the procedure manual in detail to understand the laboratory's standard operating procedures, ensure that all significant information and instructions are included, and that actual practice matches the contents of the procedure manuals.

Deficiencies detected in the procedure manual should be listed in the Inspector's summation report.

The use of inserts provided by manufacturers is not acceptable in place of a procedure manual. A manufacturer's procedure manual for an instrument or reagent system may be acceptable as a component of all the overall laboratory procedures. Any modification to or deviation from the procedure manual must be clearly documented.

Card files or similar systems that summarize key information are acceptable for use as quick reference at the work bench provided that complete manual is available for reference and the card file or similar system corresponds to the complete manual and is subject to document control.

A documented procedure manual must be developed for the tuberculosis and HIV/AIDS laboratory and be available at the work bench. Its elements should include; test principles, clinical significance, specimen type (s), required reagents, calibration, quality control, procedural steps, calculations, referral intervals and interpretation.

SAFETY PRECAUTION IN THE TUBERCULOSIS AND HIV/AIDS REFERENCE LABORATORY.

As a safety precaution, waste disposal policies and procedure must be put in place for adequate disposal of hazardous wastes. Sample handling, universal and standard precautions must be used in handling all blood, body fluid and other TB/HIV/AIDS clinical specimens.

Personnel must be very knowledgeable on the proper use of protective safety clothing and equipment such as gloves, gowns, masks, eye protectors etc.

There must be laid down procedures detailing sample transportation and handling of all patient specimen to ensure that all specimens are submitted in appropriately labeled and well –constructed container with a secure lid to prevent leakage during transportation.

Biosafety cabinet of appropriate levels should be used especially for assays involving isolation of highly infectious pathogens Tuberculosis and Human immune deficiency virus (TB/HIV/AIDS).

WATER REQUIREMENT IN THE TUBERCULOSIS AND HIV/AIDS REFERENCE LABORATORY.

In the use of reagent grade water , the TB/HIV/AIDS reference laboratory must define the specific type or grade of water required for each of its test procedures and should adequately supply same.

In the use of running water, the laboratory should periodically evaluate its source of water for silicates and other impurities.

Records of quality checks of water must indicate corrective action taken when tolerance levels of impurities are exceeded.

REAGENTS AND KITS IN THE TUBERCULOSIS AND HIV/AIDS REFERENCE LABORATORY.

The sensitivity of a TB or HIV diagnostic kit , reagent or assay can indicate the ability for it to detect very small amounts of analyte (e.g. Antibody) or ability of the assay to detect truly infected individuals.

The specificity of a kit or assay is its ability to identify all non-infected individuals correctly and produces no false positive results.

These are the basic standard practices required in a TB/HIV/AIDS reference laboratory that is involved in tests to back -up patient's clinical management.

CONCLUDING REMARKS

Important areas of TB/HIV/AIDS reference laboratory safety research need are diverse but the following examples are selected : There must be safety and efficacy of blood services, new methods of eliminating residual infectious agents from tested donor blood, new biochemical safety products for worldwide distribution ie., needles, incinerators etc. , update on biosafety in the TB/HIV/AIDS reference laboratories, evaluation and implementation of safer needle devices as part of re-evaluation of appropriate engineering controls during employer's annual exposure control plan , documentation of the involvement of non-managerial , frontline employees in choosing safer devices , and the establishment and maintenance of a sharp injury log for recording injuries from contaminated sharps 1 .

There is also an urgent need for the development of more potent and effective types and modes of disinfection and sterilization in the tuberculosis and HIV/AIDS reference laboratory.

CHAPTER 34

FIRE FIGHTING AND OTHER SAFETY PRECAUTIONS IN A MOLECULAR BIOLOGY CLINICAL REFERENCE LABORATORY.

BASIC PRINCIPLES OF FIRE FIGHTING AND PREVENTION IN A TB/HIV/AIDS REFERENCE LABORATORY.

This chapter examines the subject of fire prevention which has been of great significance and importance throughout man's history. This is because although laboratory scientists has learnt to use it every day for flaming , sterilization, decontamination, boiling , heating , incineration of wastes and other laboratory work, he has yet to learn how to control fire especially scientists in a tuberculosis (TB), human immune deficiency virus (HIV) and acquired immune deficiency syndrome (AIDS) reference laboratory.

If fire is not properly controlled in the TB/HIV/AIDS reference laboratory, fire can destroy both the scientist and the TB/HIV/AIDS reference laboratory.

It is therefore imperative that the knowledge of control and prevention of fire outbreak in the TB/HIV/AIDS reference laboratory must be pursued.

Fire is a rapid combination of two or more combustible substances resulting in the production of heat and usually light. This definition deserves careful evaluation. In actual fact, three elements are necessary before fire can occur and hence in the absence of one of the elements, there will be no occurrence of fire.

These three elements are heat, oxygen and fuel. They are required for consumption to take place in a TB/HIV/AIDS reference laboratory.

Heat: This is a form of energy produced by the upward change of temperature of a body.

Oxygen: This is a gas present in the air, which promotes burning.

Fuel: This is a combustible material in the form of a solid (wood, paper), liquid (paraffin, ether, alcohols) or gas (methane, propane).

This review does not aim to deal with large fires or those with special risks, but it will highlight the common causes of fire outbreaks that are very often ignored in a TB/HIV/AIDS reference laboratory.

SOME COMMON CAUSES OF FIRE IN A TB/HIV/AIDS REFERENCE LABORATORY.

Most fires begin in a TB/HIV/AIDS reference laboratory in a very small way and this can usually be readily extinguished if some simple form of extinguisher is available and used by somebody who understands its operation.

The risk of loss of life and laboratory properties through fire or personal injury and of the destruction of a large amount of valuable laboratory property can be substantially reduced if precautions are taken to prevent a fire starting and to prevent its spread in the TB/HIV/AIDS laboratory.

During fire outbreaks in a microbiology laboratory, it is sometimes difficult to obtain an accurate information as to the cause of the fires. This is because fire is not usually detected until it has spread and much damage has occurred.

At the time when fire is actively destroying materials in the laboratory, detecting its source is largely a matter of speculation. However, the following three situations must be considered as possible causes of fire outbreaks in TB/HIV/AIDS reference laboratories:

Negligence and careless handling of naked fire in the laboratory.

These include laboratory staff with matches or inflammable objects, careless disposal of smouldering cigarette butts, careless use or neglect of a lighted candle, stove or Bunsen burner (flame boy), smoking in a prohibited area of the laboratory, uncontrolled burning of bush or refuse, neglect of smouldering or unextinguished candles near curtains, paper stacks or book-shelves and lighting a burner when inflammable liquids like methylated spirit are being dispensed or used.

Faulty electrical circuits and appliances.

This fire outbreak can be caused by overloading electrical circuits by the use of multi- adaptors, installation of inappropriate electrical fuses and overheating of electrical appliances.

Improper storage of inflammables.

Inflammables should be stored in lockers at floor level instead of shelves . Cooking gas cylinders should not be stored inside the laboratory. There should be no leakage of gas fractured pipes , loose connections or improperly closed supply valves. Dispensing or decanting of inflammable liquids in the open laboratory instead of a fume cupboard should not be allowed. There should be no storage of inflammable liquids in improperly stoppered containers.

CLASSIFICATION OF FIRES IN A TB/HIV/AIDS REFERENCE LABORATORY.

There are four main classes of fires; Class A: This is fire involving free - burning materials such as paper, wood, and textile; Class B: This includes fire involving highly inflammable material such as petrol, paraffin, or kerosene; Class C; This class of fire involves gases- oxygen, methane, and hydrogen.

Class D: In this class, only metals are included.

ELECTRICAL FIRES IN A TB/HIV/AIDS REFERENCE LABORATORY.

Electrical fires are not placed into any of the above classes since any fire involving or started by faulty electrical systems may come under class A, B, C, or D. Therefore, the approach in combating electrical fires is to observe the materials that are burning and apply the appropriate fire fighting equipment.

If the nature of the materials on fire cannot be determined with certainty, special extinguishing agents that are non-conductors of electricity are appropriate. These include vapourising liquids, dry powder and carbondioxide, although the latter cooling and condensation effects may damage sensitive equipment.

APPROPRIATE EXTINGUISHING MEDIUM SUITABLE FOR EACH CLASS OF FIRE OUTBREAK IN A TB/HIV/AIDS REFERENCE LABORATORY.

CLASS OF FIRE   APPROPRIATE EXTINGUISHER

A   Water, C02,

B   Foam, dry powder extinguisher

C   C02, water, fire extinguisher

D   Specially designed dry powder fire extinguisher

IDENTIFICATION OF COLOUR AND CYLINDERS OF FIRE EXTINGUISHER IN A TB/HIV/AIDS REFERENCE LABORATORY.

Red cylinder   Water, CO 2 gas (9 litres capacity)

Blue cylinder   Dry chemical fire extinguisher   weight range 1–75kg on wheel.

Butter/yellow   Foam chemical fire extinquisher weight range 9–75 on wheels.

BlackCO 2 gas fire extinguisher weight 2kg– 30kg.

PROCEDURE FOR FIRE ROUTINE AND EVACUATION IN A TB/HIV/AIDS REFERENCE LABORATORY.

Fire has always been a serious hazard and particularly in laboratories where every year , a lot of money go up in flames. The following procedures is to be observed in the event of an outbreak:

1.   Anybody who first observes a fire must go to the nearest alarm point and operate the fire alarm.

2.   At the sound of the fire alarm, the fire service should be called by the fastest means of communication.

3.   In an evacuation procedure, workers must switch off all machinery in operation and proceed to a central assembly point by the nearest fire escape route to prevent oxygen feeding the fire.

4.   Once outside the burning building, in no circumstances should anyone attempt to return to it.

ASSEMBLY POINT IN A TB/HIV/AIDS REFERENCE LABORATORY.

Persons evacuated from the building should assemble at a point where the laboratory director or any leader of the laboratory should take a roll call to account for everyone. Thereafter, a report is prepared for submission to the fire officer.

A fire party or safety squad will mobilize themselves to attack the fire with all available equipment without exposing themselves to undue risk , pending the arrival of the fire brigade.

On the arrival of the fire brigade, the laboratory safety officer will report to the fire officer in charge of any fire appliances in the building about the situation of the personnel and the fire.

WHAT TO DO IN CASE OF FIRE OUTBREAK IN A TB/HIV/AIDS REFERENCE LABORATORY.

1.   When a potential outbreak of fire is detected, an alarm must be raised.

2.   Close windows and doors to prevent draught which can promote the spread of flames in the laboratory building.

3.   Ensure that all electrical appliances are switched off immediately and plugs pulled from the sockets.

4.   All personnel must be evacuated from the laboratory and no one should be allowed to go back. 1

The fire service should be called immediately on 999. In placing the distress call , the caller's name and the location of the fire outbreak must be accurately given.

If water is available , flames must be doused with ample water to prevent the spread before the fire service personnel arrives.

If clothes or laboratory coats catch fire , the victim must be laid on the floor and rolled in blankets , rugs or thick coat. Standing upright or running with burning clothes will fan the flames and result in more severe burns.

METHODS OF EXTINGUISHING FIRES IN A TB/HIV/AIDS REFERENCE LABORATORY.

1.   **Cooling method**: Rise in temperature to the point of ignition of materials is the major factor in combustion. Any agent applied to reduce temperature below the ignition point will prevent fire from burning. Therefore among the primary formula in fire extinguishing is cooling, by which water may be used to bring down the temperature of burning material below ignition temperature.

2.   **Smoothering method**: This method is used to cut off oxygen from feeding the fire. Oxygen exclusion can be achieved by smothering burning materials with fire blankets or covering with sand.

3.   **Starvation method** : This method involves manual removal of burning material into a separate place of safety. This is because as long as combustible material is available , fire will continue to burn.

The principles of fire extinguishing in a TB/HIV/AIDS reference laboratory are cooling, smothering and starvation.

FIRE SAFETY AWARENESS IN THE TB/HIV/AIDS REFERENCE LABORATORY.

In the TB/HIV/AIDS reference laboratories, , the media room is the most common area for fire outbreaks. In the TB/HIV/AIDS reference laboratories, fire occurs almost daily in the media preparation rooms.

Flaming under the biosafety cabinet can also be a source of fire outbreak if methylated spirit is mistakenly spread under the biosafety cabinet when Bunsen burner is lighted.

Among other reasons, fire could break out following gas explosion in the media room. On other occasions , waste paper left near the fire or dry napkins used for removing hot crucibles and other laboratory utensils from the burners and hot plates may cause the fire outbreak.

Quite often, the lighted end of a cigarette stick or a smouldering matchstick tossed carelessly into the laboratory dustbin may be the starting event of a fire outbreak in a TB/HIV/AIDS reference laboratory.

CONCLUDING REMARKS.

From the foregoing, it is therefore advisable to be fire safety conscious while working in the TB/HIV/AIDS reference laboratory. Therefore extra precautions must be taken to refrain from lighting a burner when the air is filled with gas .

Gas cylinders in the TB/HIV/AIDS reference laboratories should not be refilled while the bunsen burner is lighted on. TB/HIV/AIDS reference laboratories should be equipped with fire extinguishers.

www.ingramcontent.com/pod-product-compliance
Lightning Source LLC
Chambersburg PA
CBHW030933180526
45163CB00002B/553